H Y D R A

AND THE BIRTH OF
EXPERIMENTAL BIOLOGY — 1744

A B R A H A M   T R E M B L E Y ' S

MEMOIRS CONCERNING THE NATURAL HISTORY OF A TYPE OF FRESHWATER POLYP WITH ARMS SHAPED LIKE HORNS

Sylvia G. Lenhoff and Howard M. Lenhoff

A. TREMBLEY

né a Genève le 3 sept. 1710

This portrait was published as the frontispiece of *Instructions d'un Père à ses Enfans, sur le Principe de la* Vertu et du Bonheur, by Abraham Trembley, Chirol, Geneva, 1783. It is signed: Clemens del: & Sculps. 1778.

HYDRA

AND THE

BIRTH OF EXPERIMENTAL BIOLOGY
—1744

ABRAHAM TREMBLEY'S
MÉMOIRES CONCERNING THE POLYPS

By *SYLVIA G. LENHOFF* and *HOWARD M. LENHOFF*

BOOK I

SOME REFLECTIONS
ON ABRAHAM TREMBLEY AND
HIS *MÉMOIRES*

BOOK II

A TRANSLATION FROM THE FRENCH
OF
*MÉMOIRES, POUR SERVIR À L'HISTOIRE
D'UN GENRE DE POLYPES D'EAU DOUCE,
À BRAS EN FORME DE CORNES*

Distributed
by

The Boxwood Press
183 Ocean View Blvd.
Pacific Grove, CA 93950

408—375-9110

ISBN: 0-940168-01-4

Library of Congress Cataloging-in-Publication Data
Lenhoff, Sylvia G.
    Hydra and the birth of experimental biology—1744.

    Bibliography: p.
    Includes index.
    Contents: bk. I. Some reflections on Abraham Trembley and his Mémoires—bk. II. A translation
from the French of Mémoires, pour servir à l'histoire d'un genre de polypes d'eau douce, à bras en
forme de cornes.
    1. Hydra. 2. Biology, Experimental—History. 3. Trembley, Abraham. 4. Biologists—Genevan—
Biography. I. Lenhoff, Howard M. II. Trembley, Abraham, 1710-1784. Mémoires pour servir à
l'histoire d'un genre de polypes d'eau douce, à bras en forme de cornes. English. 1986. III. Title.

QL377.H9L427   1986   593.7'l   86-8231

Printed in U.S.A.

DEDICATION

"QUITE often I have had occasion to think that the care father took with our education is a truer source of happiness for us than the riches other fathers leave their children."

Abraham Trembley,
in a letter to his brother, Jean,
December 1, 1744 (Trembley, 1744*d*).

TO OUR PARENTS:

Dora and Hyman Grossman
and
Goldy and Charles Lenhoff

WHO, like the forebears of Abraham Trembley,
fled religious persecution and made many sacrifices that enabled
their children to pursue an education in a free society.

FOREWORD

Because progress in any scientific discipline depends on past discoveries and theories, one might imagine that the history of a field would be important to those working in it, and that knowledge of earlier investigators and their accomplishments would be widespread in the scientific community. That this is not the case says a great deal about how science actually works; in some respects it resembles a giant jigsaw puzzle where the filled-in parts are essential, so long as they are correct, but where it doesn't matter too much who filled them in. In practice the names of past scientists are remembered if their accomplishments are so major, like those of Darwin or Einstein, that they revolutionize our way of looking at nature. But for those just slightly further down the ladder in importance, a more likely way to be honored is to have one's name attached to a law or principle or even a piece of equipment. For every person who speaks of volts or amperes or petri plates, not one in ten knows that the memory of an earlier investigator is being invoked.

So it is not really surprising that the name of Abraham Trembley is virtually unknown among practicing biologists, and that the details of his truly revolutionary experiments on the hydra, carried out over 200 years ago, have been read by only a handful. This is not to say that the essence of Trembley's work is unknown to modern biologists. Just the contrary: we all learned in school that "primitive" animals regrow missing arms, legs, and tails; and that a few like the freshwater hydra and planaria can regenerate a whole new animal from a part. And we were taught that these same primitive organisms often reproduce by budding and fission. Indeed, just because regeneration and asexual reproduction are such fundamental biological processes, one imagines that they are self-evident facts that did not have to be discovered by a single individual who performed real experiments. That individual was, of course, Trembley, and he accomplished his major work during a short period when he lived in Holland as tutor to the children of Count Bentinck.

Shortly after his studies on the hydra became known to the European scientific community in the 1740s, Trembley experienced a brief period of fame. He was elected a Fellow of the Royal Society of London in 1743 and in the same year received its prestigious Copley Medal. The next year, 1744, saw the publication of his *Mémoires* at Leiden. But despite the unprecedented nature of his observations and the profound influence his work had on subsequent progress in biology, Trembley's name faded gradually from view. His eclipse can be credited partly to his modesty and his attitude toward his own work (especially his distaste for making grand theoretical schemes) and partly to the fact that he never again carried out extensive scientific studies.

But even for those who wanted to know more about Trembley and his work firsthand, the *Mémoires* were inaccessible from the start. Except for a pirated version published in Paris the same year as the Leiden edition, they were never reprinted, and the only translations were Goeze's German edition of 1775 and Kanaev's Russian one of 1937. Today few institutional libraries own the *Mémoires*, and the cost on the rare book market, currently $500-$1,000, prevents all but the most avid collector from acquiring a personal copy. Add to this the fact that Trembley's 18th century French poses a language barrier for most present-day scientists, and the oblivion of the *Mémoires* is virtually assured. The appearance of the Lenhoffs' translation is thus a happy occasion. For the first time since the *Mémoires* were published 240 years ago, Trembley's elegant experiments on hydras are available in English.

Fortunately for us, the Lenhoffs comprise a superb team for the job. Sylvia Lenhoff's historical perspective and her ability with language, including the nuances of the original French, are complemented by Howard Lenhoff's intimate knowledge of the hydra gained during thirty years of research on all aspects of its biology. Trembley's experiments are justly famous for their simplicity, but in several places one must interpret what Trembley saw or thought he saw in light of more recent information. Here the translator who has carried out the same observations or experiments has a unique advantage over all others. The Lenhoffs provide an introduction with background information on Trembley, his times, and the significance of his work. This is particularly useful for the general reader, who will find only one other source of such information, John R. Baker's *Abraham Trembley of Geneva*, published in 1952 and itself now difficult to come by.

Finally, this translation of Trembley is appealing because the Lenhoffs have so obviously enjoyed their task. They are fascinated by Trembley, the way he approached his work on hydras, how he described his experiments, and the cautious way he proceeded from observation to tentative conclusion and back (always) to observation. There is an enormous freshness and modernity to Trembley's work, despite the two centuries that separate him from us, and the Lenhoffs have captured this in their translation.

Joseph G. Gall
Carnegie Institution
Baltimore

AUTHORS' PREFACE

Abraham Trembley of Geneva (1710-1784), while employed as tutor-in-residence for the two children of Count Bentinck of The Hague in Holland, rocked the scientific world with his discoveries concerning the "polyp with arms shaped like horns," later designated as the genus *Hydra* by Linnaeus in 1746. Trembley's first book, which presented these findings, appeared in 1744 under the title *Mémoires, pour servir à l'histoire d'un genre de polypes d'eau douce, à bras en forme de cornes* (1744b). Among the many discoveries and experiments published in those beautiful *Mémoires* are the demonstrations that: (a) complete animals can regenerate from small cut pieces of those animals; (b) animals can reproduce asexually by budding; (c) tissue sections from two different animals of the same species can be grafted to each other; (d) the materials oozing out of the edges of cut tissue have properties that fit the definition of protoplasm as described by Dujardin one hundred years later; (e) living tissues can be stained, and those stained tissues can be used in experiments; and (f) eyeless animals can exhibit a behavioral response to light.

Trembley conducted and reported his experiments with a detail, caution, logic, and rigor rare for his time. In recognition of his accomplishments, he was elected to the Royal Society of London and in 1743 was awarded its prestigious Copley Medal, considered then to be one of the highest accolades in science. Martin Folkes (1743), president of the society, wrote to Trembley that the award was in honor of "those curious and surprising Discoveries . . . entirely unobserved in the Animal Creation, and indeed never so much as thought of, till they were brought to light, and made manifest by your diligent and exact Enquiries."

Many contemporary biologists have a vague image of Abraham Trembley as a minor eighteenth-century naturalist who discovered regeneration in hydra. That a number of biologists of intervening generations thought differently is evident in several pages of their tributes to Trembley (see Baker, 1952, pp.47-48). Cuvier, early in the nineteenth century, recognized Trembley's discoveries as having "changed . . . all the ideas that had been entertained about the physiology and anatomy of animals." Von Baer wrote in 1835 that Trembley's work "gradually but fundamentally influenced physiology . . . and thus medicine itself." And the German biologist Nussbaum wrote in 1887 that Trembley's *Mémoires* were "a classical model for a detailed biological investigation." With this first complete translation into English of the *Mémoires* we hope to reintroduce Abraham Trembley to the biologists of today as a remarkable figure in the history of biology and to provide ready access to the contents of his *Mémoires* for several other audiences.

Sylvia G. Lenhoff
Howard M. Lenhoff

Costa Mesa, California

ACKNOWLEDGEMENTS

We thank a number of individuals for very special help with various aspects of the translation: Ms. Felicia Brown and Ms. Miriam Taar, for aid with translating; Robert Cowan for translating material from Goeze's German edition of the *Mémoires*; Dr. Karl Hufbauer and Dr. Virginia Dawson, for their advice regarding relevant aspects of the history of science and for pointing out some important references; Dr. Joseph Gall, for stimulating us to complete our translation, for many useful suggestions, and for writing the Foreword; and Dr. Richard Campbell for his advice, encouragement, and constructive criticisms throughout. We also thank Dr. Aram Vartanian for sharing his extremely useful comments on the draft manuscript with us, and Dr. Bentley Glass for his helpful remarks and suggestions. Without these valued contributions this translation would have been much less worthy of Trembley's original. It goes without saying, of course, that we bear sole and complete responsibility for any errors that remain as well as for certain interpretations we present which one friendly critic found overly positivist.

An important stimulus to our work, for which we are also most grateful, has come from the direct descendants of Abraham Trembley: his two great-great-great-grandsons, the late Jean-Gustave Trembley and Dr. Jacques Trembley; Bessie Trembley, the widow of Jean-Gustave; Professor George Trembley, grandson of the late Maurice Trembley; and Alec Trembley, the son of Jean-Gustave and Bessie. They have been most helpful and hospitable to us as we researched their family archives. We are especially grateful to the late Jean-Gustave Trembley for the care that he gave to retaining and cataloguing the portion of the family archives in his possession, and for the many kindnesses he and his wife Bessie extended to us as they opened their home and their archives to us.

Other than the time that we have invested in preparing these translated *Mémoires*, the actual costs were small compared to those of many such scholarly projects. Nonetheless, without the financial support and the encouragement that we have received over the years from various sources, we could not have brought the translated *Mémoires* to publication. We gratefully acknowledge the assistance given us by the Research Society of the Sigma Xi, the University of Miami's NSF institutional fund, and the University of California, Irvine. In addition we give special thanks to the Cocos Foundation of Indianapolis, Indiana, and to the Department of Developmental and Cell Biology, University of California, Irvine. The Cocos Foundation was there with moral and financial support throughout the years whenever it was needed most. The Department of Developmental and Cell Biology provided generous support through its fine word processing staff. We are especially indebted to the late Ms. Dee Ostlin, director of that staff, for her patient, innovative, unstinting and cheerful help throughout the many, many drafts of the translation, and to her capable protégées Ms. Christine Mangold, Ms. Barbara McKinney, Ms. Patricia Weber, and Ms. Margaret Foley.

To our editor and publisher, Dr. Ralph Buchsbaum, we offer our warm thanks for his commitment to making Trembley's *Mémoires* broadly accessible to the modern reader while capturing as much as possible of the quality of the original work. Sharing our enthusiasm for the *Mémoires*, he brought to this English edition his rich store of knowledge both of the invertebrates and of fine publishing. We also thank Ms. Sue Hawthorne for typographic and production assistance.

Finally, we thank our children, relatives, and friends for their understanding and forbearance as we immersed ourselves in Trembley's world, at times neglecting those for whom we care most.

Sylvia G. Lenhoff and Howard M. Lenhoff
Department of Developmental and Cell Biology
University of California, Irvine, CA 92717

July, 1986

BOOK I

SOME REFLECTIONS

ON

ABRAHAM TREMBLEY

AND HIS

MÉMOIRES

CONTENTS

I. READERS OF THE *MÉMOIRES*

Today's Biologists

It is today's active researchers in the life sciences for whom we believe Trembley's studies will hold the greatest attraction. During the periods of interest in the study of coelenterates and other lower organisms that have ebbed and flowed in the generations since Trembley's discoveries, a number of scientists from the aforementioned Cuvier, von Baer, and Nussbaum (see p. ix) to Thomas Hunt Morgan (1901, pp. 1-2, 159, 202) and John Tyler Bonner (1952, pp. 194-195; 252-253) have turned to the *Mémoires* and acknowledged a debt to them. Interestingly, the only two complete translations of the *Mémoires* ever published, a German version by Johann Goeze in 1775, and a Russian edition by I.I. Kanaev in 1937, were both the work of scientists who studied the biology of hydras. Goeze (1791 edition, p. vii) wrote in his introduction:

I do not have to convince those who know its contents that Trembley's work is worth translating. His work has had a resounding impact not only on natural history, but also on philosophy. When I began to study these two fields, Trembley was my first and most auspicious mentor. Through him I became acquainted with the polyps. Led by his hand, I attempted to observe him and follow his example in my work.

In view of the "renaissance" since the 1950s in the use of the hydra and related coelenterates as experimental animals, a complete translation into English of this classic work seems long overdue.

Current Renaissance in Research with Hydras

The popularity of the hydra among biologists emanates not only from the relative simplicity of its structure and its graceful symmetry and beauty, but also from the ease with which it is possible to raise and maintain either large or small numbers of hydras in the laboratory. The modern science and art of "hydra husbandry" began in 1954 with publication of a quantitative examination of factors affecting the asexual reproduction of the hydra. W. Farnsworth Loomis, a physician and biochemist by training, with this work (1954) and with his pioneering experimental approaches to identifying chemical factors controlling the feeding behavior and the differentiation of gonads in the hydra, awakened many young biologists' interest in using this simple animal to investigate basic biological problems.

Today the hydra serves as a research animal for scientists in many different fields of biology. It is a favorite animal of a growing cadre of developmental and cell biologists. Other general areas of research that employ hydras concern studies of algae-host endosymbioses, of the behavior and electrophysiology of animals having simple nervous systems, and of the nature of such cellular materials or secretions as mesoglea (basement membrane) and nematocyst capsules and toxins.

A number of factors account for the hydra being a productive research animal. (1) Hydras are easy and inexpensive to grow in large numbers in the laboratory, because, when fed on the readily available larvae of brine shrimp and maintained in synthetic pond water, they can double their population every two to three days. (2) Animals grown asexually by budding are genetically alike. (3) The aqueous environment surrounding the animals can be controlled and modified by the investigator. (4) Because of its small size and lack of a hard endoskeleton, this multicellular animal is easy to dissociate into its component cells and organelles and to handle intact using many of the quantitative techniques applied to simpler systems. (5) The hydra is simply constructed at the tissue level of organization, having about ten basic cell types arranged in two concentric cell layers. (6) The hydra tissue grafts easily, and isolated cells reaggregate to eventually form intact animals. (7) The cell composition of the animal can be artificially manipulated, and hydras can be altered to exist without the pluripotent interstitial cells and those specialized cells derived from the interstitial cells, such as nerve cells, gametes, and nematocytes (see Section VIII of Lenhoff, 1983a).

The *Mémoires* as a Resource for Modern Investigators

The *Mémoires* appear to be unusual among similar classical scientific works in the extent of their continued factual validity and the provocative questions they continue to evoke. As Kanaev (1969, pp. 5-6) put it, Trembley "established a number of characteristics of . . . [the hydra's] morphology and physiology so correctly that up until now his descriptions remain in science." Some of Trembley's studies of phenomena of regulation in hydras, Kanaev continued, "were insufficiently evaluated not only in the 18th century, but also in the 19th, and only in our times have they begun to be understood properly . . . ." Virtually all of the experiments described in the *Mémoires* have been repeated and confirmed over the past two centuries and as Kanaev suggests, can still serve today as models for much research in experimental morphology and developmental biology. Today's biologists dipping into the *Mémoires* can gain insight into the nature of the hydra, the range of possible experiments that can be carried out on the animal, and also the limitations of conducting research using hydras. Trembley covers virtually every aspect of the biology of the animal including its structure, behavior, physiology, development, and its interaction with prey and predators. In addition he describes methods for finding hydras in nature, for characterizing the different species, and for feeding and maintaining the animals in the laboratory.

Further, although more than thirty years have elapsed since Kanaev's remarks, many of the processes taking place at the cell and tissue levels that underlie a number of Trembley's observations still remain to be elucidated. Some of these unanswered questions in biology may be freshly highlighted for those scientists who for the first time now will be able with ease to read the *Mémoires* in their entirety. Readier access to the *Mémoires* is needed if, as several scholars urge, scientists are to reexamine Trembley's studies for the light they may yet shed on modern research problems in fields as diverse as the biology of regeneration (Goss, 1969, p. 36) and medical studies of immunology and organ transplant (Rudolph, 1977, p. 61), as well as to correct erroneous interpretations of Trembley's work that have crept into the literature over the centuries (e.g., see Ewer, 1949, pp. 104-105).

For Students, Teachers, Historians, and Virtuosi

We believe that teachers and students at various educational levels also should be able to make use of the translated *Mémoires*. The young student may gain understanding of how scientists think and work, for the material is concrete and not abstruse, and it contains no technological jargon or specialized language. The structure of Trembley's experimental organism, the hydra, is simple, as are his instruments and techniques; he articulates his research questions and design plainly, providing clear and complete directions for others to repeat his experiments. Trembley tells us in the *Mémoires* that he "often witnessed how even children can begin to appreciate the pleasures of contemplating nature. To a child, nature presents a pageant which at first entertains him but then spurs his curiosity, instructs him, enchants him, moves him, and accustoms his spirit to delight in all that is most beautiful." Marie Boas (1958, p. 18), in a pamphlet published by the American Historical Association, recommended Dr. John Baker's biography of Trembley to teachers of history as a "good study of experimental biology in the eighteenth century." The English translation of the *Mémoires* should be a similarly helpful resource.

We also will be gratified should historians looking into the translated *Mémoires* be encouraged to give new attention to this engaging figure who speaks to us clearly and interestingly across the centuries, not only of science as in the *Mémoires*, but elsewhere also of morality, politics, and religion, and especially the education of young children (see Baker, 1952, pp. 188-240).

Finally we commend the *Mémoires* to the attention of modern day "virtuosi," or enthusiasts of science, who will enjoy witnessing a fine, creative mind at work on such phenomena as grafting, asexual reproduction, regeneration, phototaxis, and other aspects of animal behavior.

In the remainder of this introduction we have two major objectives. One is to offer certain of our reflections on Trembley and his times that may help our scientific readers appreciate the nature of Trembley's contributions to the development of experimental biology through the work he describes in the *Mémoires*. We do not cover here the details of Trembley's life and his work outside the *Mémoires*. One need but turn to Dr. Baker's admirable biography (1952) for a rich source of accurate information of this kind. We provide only limited background material that we believe might enhance the reader's enjoyment of the *Mémoires* themselves. Also, in order to afford some insight into the significance of Trembley's work for those readers who are not biologists, we present an overview of the science in the *Mémoires*, attempting to use language that is not highly technical.

We begin, therefore, with a section on the structure and species of Trembley's principal experimental animal, the freshwater "polyp," or hydra. Next we consider the design of the *Mémoires* as analogous to that of a modern research monograph. We offer the perceptions of a contemporary researcher regarding the nature of Trembley's scientific approach as expressed in the *Mémoires* and seek some insight into several possible influences upon his science. We glance at the life sciences in the aftermath of the *Mémoires*. A brief section on our search for materials related to Trembley is presented. There follow some paragraphs explaining our approach as translators and editors, and a glossary. Finally, after our list of references, we provide a list of errata, a guide for reading the translated *Mémoires* and a detailed outline of their contents.

II. BACKGROUND INFORMATION ON THE BIOLOGY OF HYDRAS

Overview of the Structure of the Hydra For Readers Unfamiliar with the Animal

Except for microscopic details, Trembley's description of the structure and anatomy of the hydra is quite accurate. For the benefit of the non-biologist and to facilitate the understanding of some of Trembley's experiments, we briefly describe the anatomy of the animal.

The structure of the hydra is one of the simplest among the many-celled animals. A hydra is constructed like a two-layered hollow tube with a base, or foot at one end and, at the other, a mouth surrounded by a ringlet of about six or more hollow tentacles, or arms, as Trembley calls them (Plate 5, Fig. 1). The "skin" of this hollow tube consists of two sheets of cells: a clear outer one called ectoderm and a thicker pigmented inner one called endoderm. Because Trembley was not able to distinguish between the two cell layers, he considered both layers as a single skin. He did distinguish, however, that the skin had a clear outer part and a colored inner part. Today we know that the inner part, or endoderm, contains the food vacuoles and pigment granules which give the hydra its color. We also now know that the clear outer part, or ectoderm, manufactures and contains the nematocysts ("stinging capsules"), structures unique to the phylum Cnidaria to which hydras belong. These structures, called granules by Trembley, were finally identified by a host of authors nearly 100 years after the *Mémoires* were published.

Trembley's Three Species of Polyps

Trembley's attitude toward nomenclature appears to have been one of nearly total indifference. He allowed Réaumur, with the collaboration of another colleague, Bernard de Jussieu, to name his animals "polyps," with the addition of one ecological ("freshwater") and one morphological ("arms shaped like horns") trait to distinguish them from the octopus and other "polyps of the sea." The term *Hydres* also appears a few times in the *Mémoires*, but it refers to the "monster" polyps created by Trembley in the course of his sectioning experiments. Thus Trembley did bring the term "hydra" into play for the first time to describe the freshwater coelenterate we now know by that name, but he used it with that strictly limited meaning only.

Trembley did not name the species of hydras with which he worked, referring to them as he tells us "according to the order in which I found them," their color, or the length of their arms. His clear description of the three species of hydras that he discovered and studied and the accurate drawings by his remarkable illustrator, Pierre Lyonet, however, make it easy to identify those animals by their current names.

The first of the three species discovered by Trembley, the small green one, is commonly known by the name of *Hydra viridis,* or, more correctly, *Hydra viridissima* (see Campbell, 1983). He was unaware that the green color comes from the chlorophyll present in the algae living symbiotically within the endodermal cells of specimens of that species. The green color of this first species of hydra discovered by Trembley contributed, of course, to the controversy over whether the polyps were plant or animal. Voltaire (see Guyénot, 1943, p. xxxvii) went so far as to say that "this growth called a *polyp* is much more like a carrot or an asparagus than it is like an animal."

The second species of hydra discovered by Trembley was later named *Hydra vulgaris* by Pallas and represents the "common brown, unstalked hydra," also called *Hydra attenuata.* Trembley distinguished this species from *Hydra viridissima* by its color, and from the third species by its lack of a "tail." Several brown species fit this description, however, and Trembley probably used specimens from two or three of these. Trembley's third species, the long-armed one with a distinctive narrow tail (or peduncle), is known today as *Hydra oligactis.* In addition to describing the morphological characteristics of his three species, Trembley also pointed out some subtle behavioral features of the animals, features usually recognized today only by biologists who have had much experience with various species of hydras. For example, Trembley noted correctly and astutely that the green hydra "contract rapidly [in response to external stimulation], whereas those of the other two species do so more slowly."

III. THE *MÉMOIRES*

Background

These were the three species of hydras that preoccupied Abraham Trembley as the decade of the 1740s opened. When Trembley began his studies of the freshwater hydra, he was an obscure young tutor. As an unknown in the scientific community, Trembley needed the helping hand of the illustrious French scientist Réaumur and the confirmation of his work by other recognized authorities in order for his startling discoveries to be acknowledged, published, and accepted.

It was Réaumur who carried news and demonstrations of Trembley's discoveries to the Paris Academy, and Buffon and Bentinck who first introduced his findings to the Royal Society of London. The *Philosophical Transactions* for 1742-1743 include selections from a letter by the physician J.F. Gronovius (1744, pp. 218-220) of Leiden which illustrate the initial scepticism in academic and scientific circles that greeted the unknown Trembley's findings.

This Discovery was and is very surprising to all our Virtuoso's [sic], and really not believed, until the Professors *Albinus* and *Mussenbrock* [sic] were provided with the Animals, and after having well examined this Creature, found the Prodigy of increasing itself in that wonderful Manner, very true.

One of the Gentlemen that made this Discovery was Mr. *Allemand*, a Man of great Learning and Ingenuity, Tutor to the Sons of Mr. s'*Gravensande* [sic].

There have been several of these wonderful Creatures sent to *Paris*, to Mr. *Réaumur*, from whom we hope for a particular Disertation [sic].

The Royal Society argued and withheld recognition of Trembley's work for two years after having been informed of it, asking Trembley to supply animals so that the President, Folkes, might himself repeat the experiments. Only after he had verified them did Folkes, speaking for the Society (see Trembley, 1943, p. 166), proclaim Trembley's work on regeneration to be "one of the most beautiful discoveries in natural philosophy."

During the process of verifying Trembley's findings, Folkes had shared some of the polyps sent him by Trembley with other members of the Society. One of these was the English scientist, Henry Baker. Using these and some hydra found locally, Baker repeated many of Trembley's experiments and pursued a number of interesting observations on his own. In 1743 he rushed into print a volume several hundred pages long in the form of an extended letter to Folkes, which contained both his own and some of Trembley's results.

Though both Folkes and Réaumur were angered by Baker's publication of Trembley's discoveries before Trembley had completed his promised volume on the polyps, Trembley did not show the same reaction (Trembley, 1943, pp. 190-191). It had taken several years of prodding by Réaumur, during which time Trembley was elected to the Royal Society, received the Copley Medal, and was invited to become a *Correspondant* of the Paris Academy of Sciences, before Trembley was finally prepared to publish "that lovely work," as Dr. Baker (1952, p. 40) calls it, the *Mémoires, pour servir à l'histoire d'un genre de polypes d'eau douce, à bras en forme de cornes.* The year 1744 saw the publication both of the elegant, authorized Leiden *Mémoires* by the company of Jean and Herman Verbeek and of the pirated two-volume Parisian edition by the firm of Durand, the latter version described intriguingly by Réaumur (see Trembley, 1943, pp. 195, 232) as a kind of "book for the pocket." We learn from one of Trembley's letters to Charles Bonnet that Trembley had negotiated with the Verbeeks an agreement by which he retained an unusual degree of control over details of the publication of the Leiden *Mémoires*, including choice of engraver, paper, print, and format (Trembley, 1943, p. 190).

Physically the Leiden edition of the *Mémoires* is particularly handsome. The book still retains its beauty in the quality of paper, print, and design, as well as in the figures and vignettes and other artistic embellishments found at the start and close of each section. The three artists who contributed to the *Mémoires* were indeed, in Trembley's words, "gifted." As he puts it in a tribute to them in the preface, "I was as fortunate in this respect as in my discoveries on the polyps." He calls the reader's attention to the engraved drawings on the fold-out leaves following each Memoir which help to explicate the text, particularly the last eight plates engraved by his friend and fellow investigator Pierre Lyonet, who at that time was just learning the art of engraving. Lyonet's artistry has been widely recognized for his detailed anatomical renderings of the goat-moth caterpillar, which Mees (1946, pp. 149-150) tells us "is an example of accuracy and careful observation that is thought by many good judges never to have been surpassed to this day." Lyonet's similarly fine contributions to the *Mémoires* of his colleague, Trembley, lead a commentator like Rudolph (1977, p. 53) to say that readers who have seen only the pirated two-volume Paris version minus the "beautiful engravings" of Lyonet are missing something special. The drawings used by Baker in his book on the hydra (1743), by contrast, though useful in elucidating his text, are cartoonish in style and not artistically noteworthy.

Baker's study of the polyp was published in a French edition only one year later, but there was no similar immediate translation of Trembley's *Mémoires* into English or any other language. In 1746, a self-described "Mathematical, Philosophical, and Optical Instrument-Maker" by the name of George Adams, published in his *Micrographia Illustrata,* what he called "A very particular Account of that surprising Phaenomenon, *The Fresh Water Polype,* translated from the *French* Treatise of Mr. *Trembley.*" This twenty-eight page section of the *Micrographia* is part summary, part paraphrase of particular portions of Trembley's *Mémoires* rather than a translation.

It seems likely that Trembley's Leiden *Mémoires* were published in a rather limited edition. Bonnet wrote Johann Goeze, the prospective editor of a German translation later published in 1775, that Goeze might have to be satisfied with securing a copy of the inferior Paris edition (Trembley, 1744c) since supplies of the Leiden *Mémoires* had been exhausted (see Goeze, 1791 printing, pp. XXV-XXVI). Goeze, with Trembley's approval, did proceed with his translation into German, which appeared the following year. It was 162 years later, 1937, before the only other complete translation of the *Mémoires* was published, this time in Russian by the Soviet biologist, I.I. Kanaev.

In the decades since there have been several efforts of other kinds to recognize and create greater awareness of Abraham Trembley's contributions. The year 1943 saw the publication of the Réaumur/Trembley correspondence, gathered and worked over painstakingly during a period of more than forty years by Trembley's great-great grandson, Maurice. John Baker's biography of Trembley, published in 1952, incorporated certain critical portions of the *Mémoires* in fine English translation and stimulated new interest in Trembley and the *Mémoires* among English-speaking readers.

An example of Baker's salutory influence may be found in the vicissitudes over the years of the entry on Trembley in various editions of the *Encyclopaedia Britannica*. Of the distinguished trio of friends and collaborators in eighteenth century biology—Réaumur, Trembley, and Bonnet—Trembley has received the least attention and recognition in the general literature as the years have passed. By 1911, for example, there was no separate entry on Trembley in the eleventh "scholars edition" of the *Encyclopaedia Britannica*. Bonnet and Réaumur, by contrast, received almost a page each. The *Britannica* gave Trembley only two brief references then: One, under the title "Evolution," identified him as a worker on the lower forms and a member of the Réaumur-Bonnet-Trembley trio. The other mention was merely a dismissal of Trembley's findings as secondary to those of de Jussieu. In a subsequent edition of the *Britannica,* Trembley's name was dropped from the article on evolution and only the deprecatory reference was retained. For the *Britannica* published in 1968, however, John Baker authored an entry on Trembley, which he did with his customary enthusiasm for his subject, describing Trembley as "one of the outstanding biologists of the eighteenth century."

In December, 1984, an international symposium was held in Geneva to honor Trembley on the bicentennial of his death. The symposium proceedings were dedicated to John Baker and were published in English (Lenhoff and Tardent, 1985). Still, in the United States access to the *Mémoires* themselves remains limited. The 1978 Union Catalog lists twenty-six libraries which indicated holdings of Trembley's *Mémoires*, fourteen of these being the Leiden edition. As part of what appears to be a growing trend of renewed appreciation for and interest in the work of this pioneering scientist, we are pleased to offer this first translation into English of his masterwork, the *Memoirs Concerning the Natural History of a Type of Freshwater Polyp with Arms Shaped like Horns.*

The *Mémoires* in their Genre

Trembley's *Mémoires* are in many ways characteristic of a large genre of mid-eighteenth century writings on natural history and on "insects" and "small creatures" in particular. Similar works were produced during this period, for example, by the other two members of the Réaumur-Bonnet-Trembley trio: Réaumur's multi-volume *Mémoires* on the insects, appearing over the years from 1732-1742, and Bonnet's *Insectologie* in 1745.

Mornet (1911, pp. 248-249) studied five hundred catalogues of libraries in eighteenth century Europe and tabulated the numbers of copies of the various works listed. As against Buffon's two hundred and twenty listings in library catalogues of the period examined by Mornet and the Abbé Pluche's two hundred and six, the work of the careful Réaumur is represented by eighty-two listings, Henry Baker's book by eighteen and Trembley's *Mémoires* by seventeen.

These books are the work of both professional scientists and amateurs, of academics and virtuosi. Some are pietistic in the vein of the Abbé Pluche, others iconoclastic like La Mettrie's; some enshrine the lofty theoretical debate of university faculties, whereas others exude the humbler air of the country doctor or parson during his free time examining the beauty of God's handiwork as expressed in His minute creatures. Some are multi-volume and range over a wide array of subjects, Buffon's running in encyclopedia style to forty-four quarto volumes, Réaumur's to six, and Bonnet's to eight.

Trembley may have modeled his *Mémoires* on Réaumur's classic in various elements of design and format. The *Mémoires* of both men also are similar in an insistent emphasis on careful reporting of observations and experiments, as contrasted with the heavy theoretical speculation that was still rampant in many of the treatises on natural history popular throughout the Age of Enlightenment even while those treatises included material from the rigorous scientific inquiry that some of their authors were beginning to pursue.

The period was one of prolific system making and bitter theoretical controversy among the biologically-inclined scientific savants of Europe. Ovist versus animalculist, preformationist versus epigeneticist, Cartesian versus Newtonian, mechanist versus vitalist, the supporter of the concept of spontaneous generation, of the idea of the Chain of Being, and so on. They held forth, debated, and disputed each other in the various public forums of the time, the journals, the burgeoning scientific societies, the fashionable salons, and of course, the world of books (see, for example, Mornet, 1911; Caullery, 1933, p. 31 ff.; Guyénot, 1941, pp. 209-401;

Hazard, 1946, p.184 ff.; Lovejoy, 1955, p. 66 ff.; Vartanian, 1963, p. 173 ff.; Ritterbush, 1964, p. 65 ff.; Gasking, 1970, p. 55 ff.).

The antitheoretical stance taken by Trembley in the *Mémoires*, however, was more pronounced even than that of his role model, Réaumur, who, for example, speculated on the animal soul (Réaumur, 1742, Vol. 6, p. lxvij). John Baker (1952, p. 183) asserts of Trembley that "it would be difficult to name a scientist who has pushed this objection [to theory] farther." Trembley combined the presentation of an unusual level of experimental detail with such adamant rejection of speculative generalization (in Trembley's terms, dangerous "so-called general rules") that he drew upon himself the mockery of some of the grand literary figures of the Enlightenment, including Fielding, Smollett, Voltaire, and Goldsmith. From Goldsmith (see Freedman, 1966, Vol. 1, p. 472), for example, we hear in *The Bee* of "the puny pedant, who finds one undiscovered property in the polype, or describes an unheeded process in the skeleton of a mole, and whose mind, like his microscope, perceives nature only in detail."

Trembley's opposition to theory and system building in natural history was so strenuous that it caused even Jean Trembley, his admiring nephew, and John Baker, his enthusiastic modern biographer, to concede that Trembley may have carried his aversion to theory a bit too far. Three years after his uncle's death, Jean Trembley (1787, p. 44) wrote that "perhaps his reserve on this issue was too great; perhaps the conjectures of such a precise and cautious philosopher would have given truth to new observations and opened a new field of study to naturalists."

On the other hand, from the vantage point of the biological scientist or student of hydra in the 1980s who turns to Trembley's work, the distinctive pragmatic experimentalism of the *Mémoires* is quite gratifying. Readers who may be seeking insight into the animal, detailed information about it, guidance in delicate experimental procedures, and ideas of related investigations yet to be done using modern technology, will appreciate the anti-speculative factualism of the *Mémoires*, which stands in sharp contrast to much of the writings on natural history of the period.

The *Mémoires* as a Research Monograph

It appears to us that Trembley wrote for particular audiences and with several major purposes. As he expresses it in his preface, he believed his observations on the polyps "could bring pleasure to the inquisitive and contribute something to the progress of natural history." Such expressions of purpose were common throughout his era. Enlightenment figures like Diderot and Condorcet exhorted their philosopher colleagues to perform the important duty of popularizing philosophy and science in order to hasten the progress of mankind (Mornet, 1911, p. 174; Vartanian, 1963, p. 19; Hahn, 1971, pp. 37-38).

Trembley may have been concerned that his discoveries, which were so much the vogue in the salons of Europe and so provocative of both scientific and philosophic speculation and controversy, be properly understood by the general literate public. Discoveries so contrary to generally held ideas require "the clearest proofs," he states at the outset of the first Memoir. If Trembley hoped by his careful account to restrain the excesses of reaction to his discoveries by the virtuosi and to "bring pleasure to the inquisitive," one nonetheless feels that Trembley was writing most of all to the likes of the amazed Réaumur, to Bonnet, de Jussieu, Allamand, de Villars, Folkes, Baker, Lyonet, and those many other "competent judges" of whom he often speaks. These were his colleagues in England and on the continent, whether amateur or professional, country parson or academician, who were engaged in the increasingly widespread serious study of small creatures.

One way to encourage the "progress of natural history" was for these far-flung colleagues who would read the *Mémoires* to repeat and verify his experiments and to contribute new and varied experiments of their own, both on the polyps and on other related animals. To this end Trembley penned his detailed "considerations" and "precautions," or in current scientific terms, "materials and methods." In his preface, Trembley tells us that from the outset, as he began to make his discoveries, he was eager to have them confirmed by others. Thus he often followed the practice of performing experiments in the presence of observers. In addition, he shared hydra from his own stocks, as well as instructions on how to carry out his experiments, with those interested in testing his experiments independently. Trembley's employer and benefactor, Count Bentinck (1744, p. 282), in a letter introducing Trembley's work to Martin Folkes, President of the Royal Society, confirms that the young Genevan made such sharing a *"Point d'honneur."*

Trembley's apparent desire to reach an audience of prospective collaborators seems to have greatly influenced the structure, format, and content of the *Mémoires*. The Leiden edition consists of a single volume as contrasted with the generally more voluminous publications on natural history then common. The conciseness of the work reflects in part the author's style and emphasis on facts, and also the nearly exclusive, one might say almost exhaustive, focus on a particular organism in contrast to the compilations of studies on a variety of creatures and concerns contained in many similar contemporary works. In the *Mémoires* Trembley does include work on related "polyps," on animals that serve as food for the polyp, and on other small creatures that appear to share some of the reproductive peculiarities of the polyp, such as the bryozoan *Lophopus* and the annelid *Stylaria*. In the preface, referring to earlier work by Bonnet and Lyonet, he devotes several pages to publishing Lyonet's further findings on the parthenogenic reproduction of aphids and urges his colleague "to publish a complete work" on his own. With these and a few other minor exceptions, the *Mémoires* essentially constitute a pioneering research monograph on a single animal, the freshwater hydra. Kanaev (1969, p. 6) describes the *Mémoires* in a similar vein as a "monograph."

Drawing upon observation and experiment already reported in detail in lengthy letters to Réaumur, to the Royal Society, and others, Trembley organized the *Mémoires* in a very methodical, topical manner. (See our outline of the *Mémoires* on p. 56). Trembley tells us in the first Memoir that he will proceed in the order "most natural to me," an order which proves to be remarkably similar to that of a modern scientific report. Memoir I is essentially an "introduction" of the polyp to the reader, with a general description of its form and movements and some observations on the structure of the animal's parts. It is noteworthy that detailed material on structure is reserved for the relevant sections on function. In Memoir II Trembley deals with "materials and methods" of the animal husbandry entailed, how to collect the polyp, feed and maintain it, with some observations on color and functional morphology that are related to its feeding. Memoir III is a "results" section, dedicated entirely to one of the most notable attributes Trembley has discovered in the animal, that is its "amazing reproduction" by asexual means. The major portion of Memoir IV completes the "methods" and "results" as Trembley presents all the other "operations" he has carried out on the polyp. These include sectioning the animal in almost every manner conceivable; the making of monsters; and the famous inversion experiment in which Trembley details for us his procedures for deftly turning the tiny creatures inside out, the experiment which led to the first experimental grafts of animal tissue.

The final pages of Memoir IV, which are set off from the experimental material, are very much analogous to the "discussion" section of a modern scientific paper. Here Trembley discusses the relationship of his polyps to the "polyps of the sea" or cuttlefish, and other presumably polyp-like creatures, presenting a "literature search" on the subject. The search is instructive regarding the progress of zoological studies to that time. Trembley dutifully incorporates references to the ancients, Aelian, Aristotle, Augustine, Massarius, and Pliny, wryly remarking, "I believe one may be allowed to doubt the accuracy" of some of their assertions. Elsewhere he demurs from judging the degree to which the ancients' views should be heeded since such judgments require knowledge of the specific observations on which their views were based and how they carried out these observations. "Such details," he says with profound understatement, "are not found in the works of any of these writers." From the ancients Trembley leaps centuries to Swammerdam, to Réaumur, and to other contemporaries such as the English minister, Mr. Hughes, who have demonstrable factual information to impart.

Proceeding to a discussion on the question of characteristics distinguishing animals from plants, Trembley permits himself to hold forth at some length on the perils of general rules, of hypothesizing on the basis of insufficient facts, and on the importance of drawing limited conclusions. Polyps, he says, do not constitute some newly hypothesized class of "zoophytes" or "animal-plants," as he himself had once suggested (Trembley, 1943, p. 61); they should be looked upon instead as simple animals. He urges philosophers to drop the preconceptions that blind them so that they can pay attention to the facts before their eyes, just as children do. He argues that had men not been held back by suppositions that one or another thing was "impossible," natural history would be far more advanced and regeneration among animals, for example, would have been discovered long since. (Disarmingly, Trembley admits that his own supposition of this kind, that pieces of an animal could not become complete animals, contributed significantly to his discovery of regeneration.) He ends this final "discussion" section and the *Mémoires* with a plea for expanding careful observation and experiment and not mixing our own notions with what we learn from closely examining nature itself.

In one way the *Mémoires*, despite their organization, factualism, and experimentalism, do not read like a modern scientific report. Trembley, the scientist, with his emotional reactions, his notions conjured with and dismissed, and reflections and elements of his personality—these remain to enhance the pleasure of reading the science itself. For example, when Trembley first observes hydras that are feeding, he comments: "Only reluctantly did I absent myself for a few hours from this spectacle which had so greatly excited my curiosity. Impatience to know what would become of the millepede drew me back to my study as soon as possible." He reacts with similar pleasure and excitement to his first observation of asexual reproduction by budding in the creature and to his discovery of the regenerative powers of pieces of the tiny animal.

Major Scientific Contributions Described in the *Mémoires*

Most references to Abraham Trembley describe him as the discoverer of regeneration. Few authors, however, with Baker (1952) being one of the exceptions, note the range of discoveries made by Trembley using hydra as reported in the *Mémoires*. Still fewer scientists and historians appear to be aware of other major findings by Trembley that are not mentioned in the *Mémoires*, such as the first description of dividing cells. This observation, which he first described in a letter to his former employer Count Bentinck, was made on the single-celled diatom known today as *Synedra*. These observations were later announced in a publication by Bonnet in 1765, and in 1775 by Trembley in one of his books on the education of children (see Baker, 1952, pp. 155-158).

In Table 1, we list nearly 60 original contributions made by Trembley as he worked with hydras, including discoveries that add much to our knowledge of the animal. From the headings, it can be seen that those discoveries cover a broad range of disciplines. We have placed asterisks next to Trembley's major findings with hydras, most of which were subsequently determined to be significant in regard to other animals as well. In the table, asterisks also mark important tenets of the

Table 1. Scientific contributions to be found in the *Mémoires*

A. Scientific Philosophy	E. Developmental Biology (cont.)	G. Physiology (cont.)
*1. Distrust general rules *2. Operationalism—give detailed methods *3. Exercise caution—withhold judgment *4. Do more and varied experiments *5. Repeat experiments frequently *6. Repeat under natural conditions *7. Let the organism guide the direction of your research	*4. Inversion (reversal) experiments *5. Grafting experiments *6. Relationship of amount of food consumed and temperature to budding rate 7. Developmental abnormalities, naturally occurring and induced 8. Polarity of regeneration 9. Multiplication by transverse fission 10. Continuing development of bud when half of parent removed 11. Healing of wounds of cut hydras	7. Transfer of food from gut to buds and to arms *8. Ability of hydras to recognize different or same species in grafting experiments and attempts at forced cannibalism
B. Natural History		**H. Research on Other Organisms**
1. Observing under natural conditions 2. Finding hydras in nature 3. Noticing seasonal variation in distribution and number of hydras		1. Various prey animals, such as *Daphnia* and *Tubifex* 2. Body lice (*Kerona*) on hydras *3. Anatomy of *Lophopus* (a bryozoan) *4. Budding by *Lophopus* *5. Budding by *Stylaria* (an annelid)
C. Taxonomy	**F. Behavior**	**I. Ecology**
*1. Characterized morphological differences among 3 kinds of hydras 2. Described behavioral characteristics of different species 3. Observed differences in pigmentation among the species	1. Activities of parts of body 2. Stimuli for contractions 3. Different kinds of locomotion 4. Feeding behavior *5. Propensity for light 6. Mechanisms for suspending from surfaces of water 7. Separation of buds *8. Observation of habituation to contractions stimulated by mechanical agitations 9. Development of feeding behavior of developing bud 10. Observation that well-fed animals do not readily eat *11. Experiment indicating hydras have memory of light (see Josephson, 1985)	1. Hydras move to sites where food is most abundant 2. Some types of waters support growth of hydras, whereas others kill the animals 3. Hydras in nature vary in numbers depending upon the season 4. Hydras attached to snails advance more rapidly
D. Morphology		**J. Techniques and Methodology**
1. General structure of hydras 2. Number and distribution of arms 3. Structure of body 4. Structure of foot 5. Structure of buds 6. Pigmentation of hydras 7. Structure of area around mouth 8. Description of structures shown later to be spermaries and eggs 9. Description of clusters of granules in tentacles shown 100 years later to be nematocysts		1. Methods for finding hydras 2. Best waters for culturing hydras *3. Use of colored foods as vital stains to study uptake of nutrients, and paths of distribution of food 4. Method for observing an extended arm 5. Methods for measuring budding rates *6. Method for inverting hydras *7. Method for distinguishing between attraction to light or temperature or air *8. Method for making many-headed "Hydras" 9. Methods for collecting live food for hydras
E. Developmental Biology	**G. Physiology**	
*1. Development of buds *2. Regeneration experiments *3. Budding experiments; rate of budding	1. Factors affecting contraction and elongation 2. Effect of temperature on many body processes 3. Uptake of pigments and nutrients 4. Digestion and egestion *5. Properties of viscous material (protoplasm) 6. Adhesion to surfaces	

scientific philosophy (category A) that Trembley espoused in the *Mémoires*, as well as some of the unique or generally applicable techniques (category J) that he described for the first time. All of these findings and techniques were developed using hydra, except those discussed in category H.

Among these various contributions by Trembley, those basic discoveries in developmental biology listed in section E of Table 1 probably are the best recognized. He was the first person to show that animals can be made to reproduce by cutting them into separate pieces and allowing those pieces to regenerate (E 2). He also was the first to demonstrate conclusively by experimentation that an animal can reproduce asexually by budding (E 1, 3, and 6), and he extended those observations made on hydra to a bryozoan (H 4) and an annelid (H 5). In addition, Trembley is credited with carrying out the first true grafts made with animal tissues (E 5). Finally, his inversion experiments (E 4), called reversal experiments by some, stimulated debate for years to come about the fate of the ectodermal and endodermal layers (see Baker, 1952, pp. 74-77).

In the field of animal behavior, Trembley has more recently been accorded greater recognition (Bodemer, 1967) as the first to observe and to devise experiments demonstrating that animals which do not possess eyes are attracted to and move towards light (F 5). One fascinating brief observation on animal behavior mentioned by Trembley in the *Mémoires* has not been noted previously. This is his description of habituation, i.e., the process whereby an organism stops responding to a stimulation once that stimulation has been applied a number of times in succession. We refer to the legend of Figure 1, Plate 10 of Memoir III. The figure shows "an aquatic caddis worm . . . swimming with eight long-armed polyps attached to its case by their posterior ends." In the legend Trembley notes that he has "seen a number of [these] polyps which were not induced to contract by the [swimming] motion of the caddis worm any more than those shown in this figure." Thus, Trembley, with his eye for the unusual, observed and first recorded an instance of habituation. Previously he had noted many times that hydras invariably contracted in response to mechanical stimulation (F 2). Apparently those hydras attached to the case of the swimming caddis worm eventually habituated to the repeated mechanical stimuli resulting from those motions. Such habituation was eventually described in the hydra in greater detail over two hundred years later by Rushforth et al. (1963).

IV. A TWENTIETH CENTURY PERSPECTIVE ON TREMBLEY'S APPROACHES TO BIOLOGY

Areas of Excellence

Trembley emerges in the pages of the *Mémoires* as a creative antitheoretical experimentalist who gave much attention to reporting the details of his procedures. His masterwork reveals a "laboratory personality" that combined great attention to accuracy with ingenuity and intellectual playfulness.

To our minds Trembley ranks among the leading eighteenth century biologists on a number of counts. He may be acknowledged for his excellence as (1) an *observer*, who was quick to notice the unusual and to report his findings with great accuracy and detail; (2) a *naturalist*, who discovered a number of new species; (3) an *investigator of processes*, such as asexual reproduction by budding, when many of his contemporaries were concerned primarily with describing structures and reporting events and with the philosophical problems of biology; (4) an *experimentalist* who was not content until he could prove his findings in a number of ways; (5) a *technician par excellence* who carried out complex and delicate operations, many of them in a drop of water held in the palm of his hand, using hardly more than a scissors and a boar's hair; (6) a *mathematically adept biologist* with a grasp of the importance of quantitative elements in the study of natural history who backed many of his experiments with numbers and who often repeated those experiments numerous times until he was convinced of their veracity; (7) an *organismic biologist* who investigated many phases of the life history of one animal; and (8) an *operationalist* in the sense of his belief that experiments have no lasting value unless complete directions are given regarding the methods both by which the experiments were carried out and the results observed. It is not surprising that Trembley has been called by some "the father of experimental zoology" (see Baker, 1952, pp. 171-172).

In this introduction, we will not belabor the evidence that Trembley was a first-rate observer, naturalist, investigator of processes, experimentalist, and technician. There are sufficient examples in the *Mémoires* themselves. On the other hand, we do wish to call attention to what we find challenging to think of as his organismic, operationalist, and quantitative approaches. We feel that these very features which distinguish Trembley from so many of the early biologists, at the same time may have contributed somewhat to his having faded into relative obscurity.

Trembley as a Student of Quantitative Biological Inquiry

The *Mémoires* show Trembley not only to have been an exemplary experimentalist and observer, but also to have come close to developing a form of quantitative biological inquiry. One of the better examples of the latter can be taken from examining his data on the budding rate of individual hydras (p. 106).

On those pages he presents a table which lists the number of buds observed and the order in which they both emanated and then detached from a single hydra over a two month period. From this experiment and others, he concluded that a single polyp produces an average of about 20 buds per summer month, some animals producing more and some less. So precise were his records that we have been able to plot the data and obtain the graph of the rate at which buds appeared on and separated from a single hydra (Fig. 1). The graph is virtually the same as any biologist would obtain today were he or she to conduct the same experiment. Interestingly, the graph shows that whereas the budding rate during July was about 20 buds per month, the rate during August was closer to 30 per month. Possibly the latter rate reflected the higher temperatures during August, a factor which Trembley said would increase the rate at which buds are produced.

Trembley also calculated that it takes about five days for a well-fed, newly

Fig. 1. Rate at which buds appear on and separate from a single mother.

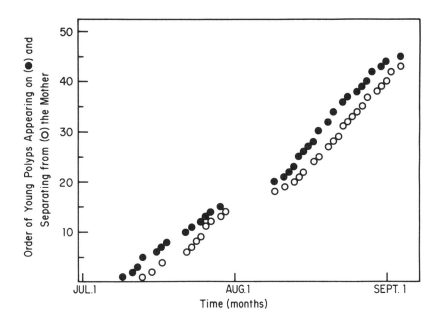

The data are taken from p. 106 of Memoir III. Apparently Trembley did not observe the exact days of appearance and separation of the buds around the first week of August. Nonetheless the buds that separated (numbers 15-18) remained in the vessel, and he was able to record this number and to continue his experiment. See our text for explanation of different rates in July and August and for an analysis of "separation times."

detached bud to begin to reproduce by budding. Hence, he concluded, "Taking all of this into consideration, it is easy to understand that at the end of two months the number of descendants from a single polyp can be prodigious." Trembley was so right. Our colleague Dr. Richard Campbell, using a computer, took Trembley's two assumptions, one, that an animal can produce 20 buds a month, and, two, that a new bud can initiate its own budding within five days, and came up with a table which is given below. The higher numbers in the second column are rounded off.

Table 2. Number of progeny developing from a single newly detached bud

Days since experiment began	Number of possible detached progeny
1	1
15	14
30	349
60	250,000
90	185,000,000
120	134,000,000,000

Although Trembley's expression "prodigious" is not precise, it certainly is appropriate. Trembley seems to hint at the exponential pattern of the multiplication of animals. His thinking in this regard may show some influence of his mathematical training in calculus (see Buscaglia, 1985).

Elsewhere in the *Mémoires* Trembley's regard for quantification is evident. For example, he tells the reader that he has repeated a particular experiment a certain number of times and with what results. When looking for the opening between a dissected portion of the parent and the bud, he writes, "I was not satisfied with doing this experiment once; I tried it on seven occasions, and succeeded on five." A little later he states that he has seen hydras divide on their own by a sort of transverse fission, but he adds the the qualifying statement, "Although I have studied a considerable number of polyps over a period of three years, I have not seen more than twelve divide [by transverse fission] on their own."

Examples of Trembley's interests in quantitative data also can be found in his discussions of the effect of temperature on both the rate and amount of food consumed, on the initiating of budding, on the detachment of the buds, and on the hydra's general ability to respond to stimuli. From a journal of his experiments and observations, he incorporates in the *Mémoires* excerpts that often include dates, the time of day of the observation or operation, and the number of animals studied.

Trembley, An Organismic Biologist

Trembley, at least in these *Mémoires*, shows himself to be a model "organismic biologist." We define an organismic biologist as one whose research is focused primarily on a single whole organism, and who investigates virtually the entire range of life processes of that organism. We contrast an organismic biologist with a problem-oriented biologist, that is one who uses an organism, or a group of organisms, in order to investigate a particular question. Once an organismic biologist starts to conduct research on an animal, that individual may be led on to investigate one phenomenon after another with no immediately apparent connection between them and without regard to the specialized discipline of biology in which he or she may have been trained. That is, at one point the subject under investigation may be behavior; at another juncture, developmental biology; at another, physiology; at still another, ecology; and then maybe back to behavior, or again to physiology. Or, as Trembley wrote in his preface, "I was swept along, as it were, from one observation to another with barely the time to make notes in my journal."

If we focus only on Trembley's experimentation, that is, not his descriptive and procedural work, we can follow the path of his experiments and see how one separate line of research grows out of a seemingly unrelated one (see Fig. 2).

Trembley's observation that hydras had a propensity to move toward light stimulated him to start investigating the animal seriously. Once he began to give the hydra all of his attention, he observed that the animals did not have an equal number of arms. Hence, he thought of sectioning the hydra into two parts just to check the vague possibility raised once again in his mind that the hydra might be a plant. This experiment led him to his elegant series of experiments on regeneration (see also page 38 of this Introduction).

When he cut the hydra into pieces, Trembley noticed that many granules from the body wall ("skin") of the animal were released into the surrounding solution. It was his recollection of this observation, in fact, when he was concerned with the color of the hydra, that led him literally from one observation to another. By examining the granules, he noticed that they were held together by "viscous material" that today we would call protoplasm. By further examining the color of the granules he got his ideas about how the hydra assimilated food along with the colored material from the prey. Because he noticed this role of the colored granules and vesicles in the lining of the stomach in taking up food, and that the outer part of the skin (ectoderm) also had vesicles, he devised a number of experiments in which he sought to nourish a hydra by placing it in a solution of nutrients. In his final experiment, he "thought of inverting them so that the external surface of their skin would form the walls of their stomach." (For an interesting interpretation of how the ideas of the great Dutch scientist Boerhaave may have influenced Trembley's inversion, or "reversal" experiments, see Dawson, 1985.) Once Trembley had succeeded in inverting the hydra, he observed in one instance that a bud had grafted on to the parent. From then on he proceeded to devise a series of experiments showing conclusively that it was possible to graft pieces from two different hydras together (H.M. Lenhoff and S.G. Lenhoff, 1984).

A flow chart of this organismic chain of experiments and observations (Fig. 2) might look somewhat like this:

Fig. 2. Organismic chain of Trembley's experiments and observations

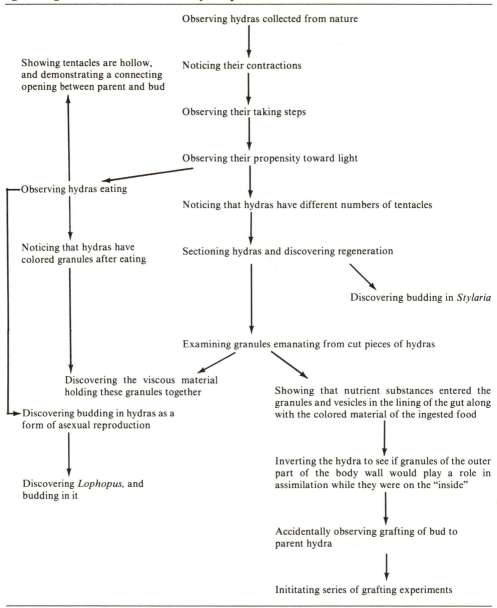

It would be disingenuous of us not to mention that the organismic approach which we ascribe to Abraham Trembley is the same one that has been used by one of us (H.M.L.) for the past 25 years in his research on the hydra. For example, he has investigated such problems as culture of hydras in the laboratory (Lenhoff and Brown, 1970), migration of cnidoblasts (Lenhoff, 1959), composition of nematocyst capsules and toxins (Lenhoff et al., 1957; Blanquet and Lenhoff, 1966; Hessinger and Lenhoff, 1976), chemical control of feeding behavior (Lenhoff, 1969; Lenhoff, 1981), mechanism of protein digestion (Lenhoff, 1961), chemical nature of endosymbiosis (Muscatine and Lenhoff, 1961), induction of budding in developmental mutants (Lenhoff, 1965; Novak and Lenhoff, 1981), composition and role of hydra's acellular mesolamellae (Barzansky and Lenhoff, 1974), hydra's pigments (Krinsky and Lenhoff, 1965), control of differentiation of gonads (Rutherford et al., 1983), use of hydras as a biological control for mosquito larvae Lenhoff, 1978), and now the history of how research on hydras began (Lenhoff, 1980; Lenhoff and Lenhoff, 1984).

Perhaps this affinity for organismic biology may account in part for our fascination with the *Mémoires*. We imagine that had Abraham Trembley discovered the hydra in 1980, he would have followed the same organismic approach that he practiced in 1740, but using such tools of the day as electron microscopy, radioisotopes, biochemical procedures, and monoclonal antibodies.

Trembley recognized this chain of organismic experimentation which led to his famous discovery of regeneration and modestly commented with regard to it: "Because of its nature, that finding was to be not the fruit of long patience and great wisdom, but a gift of chance." If there are lessons in Trembley's organismic approach for today's aspiring biologist, it might be: Do not overspecialize. Get a good background in experimental techniques, start to observe and investigate your organism, and let it—not your preconceived ideas—be your guide.

Trembley, an Operationalist

It is not enough to say . . . that one has seen such and such a thing . . . unless at the same time, the observer indicates how it was seen, and unless he puts his readers in a position to evaluate the manner in which the reported facts were observed . . . . Insofar as I am able, I shall bring the reader into my study, have him follow my observations, and demonstrate before his eyes the methods I used to make them.

These simple words, taken from the first two paragraphs of the first Memoir, state a most important aspect of the scientific philosophy of Abraham Trembley. Today we might call such an approach "operationalism," a version of the scientific method which insists that the lasting validity of an investigation resides in an accurate description of the results, the methods by which the experiments were carried out, and the means by which the results were observed; all else is considered conjecture and might not stand the test of time as new means for investigating the same phenomena are developed and new facts are uncovered.

Trembley insists that to judge the validity of an observation or experiment, one must know how it was performed and under what conditions. Hence, the *Mémoires* are full of experimental detail. The figures at the close of each Memoir, to which Trembley refers heavily, are prominently displayed and are preceded in each case by pages of explanatory notes. The descriptions of his methods in the text are extremely clear and complete, leaving little to the imagination for others who wish to repeat his experiments.

Trembley emphasized the importance of observing living organisms under natural conditions and of differentiating between experimental results obtained in the study and events observed in the animal's natural habitat. For example, in Memoir III, Trembley recounts his excitement upon discovering in one of the ditches at Sorgvliet at a certain season huge branches completely covered with polyps. Off he dashed with a sample to put it safely away in his study. He then hastened back to the ditch just as quickly to set a plank out over the water on which he could lie to observe this striking new abundance of polyps in their natural setting. As another example, Trembley shows sensitivity to the important differences that may exist in artificial as opposed to natural settings when he points out differences in the numbers of hydra tentacles he has seen in nature versus those of animals kept in his study. He comments also about the effect on the polyp's fecundity deriving from the quantity of food material available to the animals kept in his jars versus the lesser amount generally available under natural conditions.

Not only is it important to observe the animal repeatedly and in its natural environment, Trembley warns, but also under comparable conditions and at approximately the same times during the year. One reason he gives for delaying publication of the *Mémoires* was his need to repeat observations a year later so that he might carry them out under "identical circumstances."

The extraordinary virtuosity of Trembley's experiments has been noted by a number of commentators. Those experiments were also usually both comprehensive and exhaustive, squeezing out virtually all the information on a particular phenomenon that Trembley's level of instrumentation would allow (see Miall, 1912, p. 282; Bodemer, 1964, p. 21). Those instruments were limited indeed. Trembley relied more on a simple, powerful lens than on any other tool, though he did devise clever contraptions to best utilize for his purposes the microscopes available to him (see Archinard, 1985; Baker, 1952, p. 171). As Ritterbush (1964, p. 104) points out, the microscopes of the time had such low resolving power that "speculators were able to imagine they had seen whatever they liked."

Other "instruments" Trembley used extensively were artists' brushes and hog and boar bristles. But how he used them! Modern practitioners may envy his dexterity in turning a hydra inside out using little more than a hog bristle, or his ease in suspending a hydra at the surface of the water in his powder jars. The *Mémoires* show his strengths in conceptualizing pragmatic solutions to scientific problems in terms of ingenious experiments. The deceptively simple design of the famous chevron experiment is illustrative. A cardboard muff over his powder jar, with a chevron-shaped opening cut into the cardboard, was the extent of the instrumentation Trembley needed to carry out the first demonstration of phototaxis in eyeless animals. In his fine article on eighteenth and nineteenth century ideas on light and animal behavior, Bodemer (1967, p. 136) calls Trembley's chevron experiments "a paradigm of simple experimental design and cautious interpretation." He says further that "Trembley's investigation of heliotropism in the polyp . . . represents the most significant study of the influence of light upon animal behavior prior to the last decades of the nineteenth century." All this with cardboard and scissors.

Trembley as a Cautious Interpreter of Results: A Biology of Limited Conclusions and Suspended Judgments

When Trembley reported his results, a number of which were revolutionary, he couched them in cautious terms. We provide but a few examples of his caution, one or two from each Memoir. In Memoir I, when explaining the possible adhesion of the polyp by a combination of the meshing of the skin with an irregular surface, and the involvement of a viscous substance, Trembley adds, "I would not wish to allege, nonetheless, that no other causes may be involved." On this point, Trembley admitted the limits on discovery imposed by the nature of the organism and by the state of eighteenth century technology. He gave up on trying to discover how the hydra's adhesion to surfaces was controlled. "The polyp is too small an animal to permit experiments to be made that would answer this question conclusively." Similarly, Trembley points out that though he finds only one canal in the polyp, "It may be that there are some others . . . which may be so small that they have escaped my scrutiny."

In Memoir II, after Trembley has proven that hydras move toward light, he admits that he cannot find "any part which, by its location or by its structure, gave me reason to suspect that it was an eye." But does he conclude that they "have no means of perceiving light on the objects it renders visible"? No, Trembley concludes, "When facts are lacking in such research, it is more appropriate to suspend judgment rather than make decisions which almost always are based on the presumption that nature is as limited as the faculties of those who study her."

In a choice passage in the third Memoir as Trembley describes his experiment on the possible existence of a connecting opening in the polyps between mother and young, he admonishes his reader regarding the value of repeating experiments, urging that one not "become disheartened by want of success, but . . . try anew whatever has failed. It is even good to repeat successful experiments a number of times. All that it is possible to see is not discovered, and often cannot be discovered, the first time."

On this same subject, in order to show that a bud actually developed from an evagination of the body wall of the parent hydra, Trembley cut out that portion of the wall (Plate 8, Fig. 5) and saw clearly the hole connecting the gut of the parent hydra and its bud. But Trembley remained cautious in drawing final conclusions, because, as he said, "It was still possible, however, that at the place where the two polyps joined there could be a [transparent membrane] . . . which separated the two stomachs." He then proceeded to prove that there was no such membrane by observing colored food go from the stomach of the parent into that of the bud, and vice versa.

Further on in this same Memoir, when Trembley describes how he attempted to determine if the bud received some sort of "reproductive factor" from its parent hydra by means of the external environment but was unable to find any such interaction at all, he nonetheless concluded that "what I am attempting to discover, supposing something of the kind ever existed, was either imperceptible or at the least very difficult to see."

Or, in Memoir IV, after stating his findings that virtually every part of a hydra could regenerate a complete animal, but that pieces of isolated tentacles could not, he wrote, "The experiment did not succeed. I would not wish to conclude, however, that successful regeneration from a single arm is impossible." Further along in Memoir IV, when discussing his inability to graft pieces of different species of hydras together, he decided his experimentation was done "neither with sufficient care nor with sufficient frequency, however, to assert that it cannot succeed."

In the examples just given, we see Trembley practicing a science of limited conclusions; he voices the need for such an approach repeatedly in the *Mémoires*. Trembley was willing to accept that in many cases he would find no satisfactory answers despite long and laborious efforts. His attempt to understand how hydras digested their food is a case in point: "I have never flattered myself that I have acquired very precise ideas on the subject," he says. As to how the nutritious elements of the food are finally absorbed in the body walls of the hydra, he cautions, "I will not promise satisfactory answers . . . . I am simply going to set forth some observations." Finally, regarding how the nutritive material spreads from "the granules" into which it has passed, to other parts of the body, "I find myself completely unable to answer that question."

With counterpoint from his arguments for limited conclusions and suspended judgments, Trembley's antipathy to "general rules" runs like a leitmotif throughout the *Mémoires*. For example, he warns the reader not to be seduced by the "general rules" which would have the polyps be "neither plant nor animal and yet both." Or again, Trembley points out that Bonnet's discovery of parthenogenesis in aphids has already discredited the "allegedly universal rule that there is no reproduction without copulation."

Another of Trembley's targets was the widespread practice among naturalists of arguing by analogy. Ritterbush (1964, p. 124 ff.) takes as one of his main themes in characterizing the study of natural history in the eighteenth century, the development which he calls the "Triumph of Botanical Analogy," and which he describes as particularly pernicious to the progress of the life sciences during that century. Ritterbush terms Trembley "a sound experimentalist" unusual for his time and finds him "indifferent to the idea of botanical analogy." In the *Mémoires* Trembley seems definitely antagonistic rather than simply indifferent to argumentation by analogy. It might certainly be that there were various "imperceptible parts" in the polyps, but to posit them on the basis of "simple analogy," Trembley says, is not "very satisfying." Since the polyps differ in many ways from other animals, they may also differ in regard to any such minute parts.

In the matter of argument by analogy, it is interesting to contrast Trembley with the Englishman who followed him in observing and experimenting with hydras, Henry Baker. Baker, like Trembley, emphasized careful observation and experimentation. Both wrote as modest experimentalists who loved nature and nature's God, without having great systems to propound, whether religious or scientific. But it is easy to imagine what Trembley's reactions might have been to such an argument from analogy as Baker engaged in regarding the polyp's "teeth." Baker (1743, pp. 32-33) decides the polyp must have teeth even though he cannot see them because of the way it breaks up a worm into pieces it can swallow, and because of "the violent and painful Agonies a Worm experiences the Moment it is taken hold on by the Polyp's Snout."

V. TREMBLEY'S SCIENTIFIC PERSONALITY: INTELLECTUAL PLAYFULNESS AND ACCURACY

Trembley often appears almost playful as he experiments. In one passage he tells of deliberately tangling up the arms of a long-armed polyp so badly he does not believe the polyp will be able to disentangle them without help. At another time, while trying to determine the polyps' natural enemies, he is repeatedly unsuccessful in coaxing a perch to swallow and keep down a polyp. After a while he conceives of tricking the fish by presenting it with a polyp that is itself in the process of swallowing a kind of worm much favored by the perch. A similar kind of playfulness is evident in the sectioning experiments and in those on inverting hydras.

Trembley may have been creative and playful while observing and experimenting, but we must emphasize that most of all he was painstakingly exact both in what he did and in how he later described it. John Baker (1952, p. 174) finds the "accuracy of his observations . . . perhaps the most striking feature of Trembley's work. One can read his writings today not simply for their historical interest, but to get reliable information; the student can learn biology and the history of biology at the same time." Comparing members of the "distinctive trio in XVIIIth century biology" consisting of Réaumur, Trembley, and Bonnet, Baker (1952, p. 24) judged that neither Bonnet nor Trembley "achieved distinction over so wide a field" as Réaumur, whereas Bonnet "looked farther into the future than did his friends." Trembley, however, according to Baker, "was the greatest of the three as an observer: his discoveries, though far less diverse, were more important than any made by Réaumur in biology."

VI. TREMBLEY: THE MAN AND HIS TIMES

Was this young Swiss who burst suddenly through no proper academic entry way onto the stage of mid-eighteenth century science amateur or professional? Was he a brilliant isolate or was he integrated into the larger scientific and intellectual community? What can we say briefly of some of the men and ideas influencing his contributions to science? To what extent does his work seem to reflect the scientific zeitgeist of that era? To what extent does it seem to leap ahead into the future?

The Enlightenment Setting

Trembley's findings had an extraordinary reception in a milieu in which both scholarly and popular interest in natural history was extremely high. To set the scene, we first need to recall the general cultural surge of the European Enlightenment. Society and its ideas were in ferment. Cross currents of change and reaction were sweeping Europe, affecting all aspects of mid-eighteenth century life including the pursuit of science.

When Trembley began his investigations the study of natural history was in a state of flux. Baconian empiricism of the previous century had taken a strange turn in the first quarter of the eighteenth century. Bacon had urged naturalists to forego conjecture until much more data had been amassed. Acting ostensibly on this Baconian principle, many of "the curious" and professional naturalists as well had naively set out to collect great cabinets full of specimens, or as Ritterbush (1964, p. 62) put it, "torrents of shells, fossils, insects, dried plants, and other curiosities." Responding in part to this collection mania, the satirists went on the attack. In Addison's *Tatler* essays of 1710, we find that "standard comic figure of a virtuoso," Sir Nicholas Gimcrack, and in Swift's *Gulliver's Travels* (1728), "witless scientific investigation" is taken to task. Later, John Hill launched a protracted battle with the Royal Society, charging it with publishing in the *Philosophical Transactions* "many trivial and foolish articles" (see Ritterbush, 1964, pp. 61-63; Stimson, 1948, p. 70 ff., p. 127 ff., pp. 140-141). The satirists often did not differentiate, of course, between the truly trivial and the painstaking new work which gave careful attention to the "minute creation," and which was beginning to lay the foundations of modern biology through the studies by the "great observers" of the period. Ritterbush states that despite the tremendous impact Newtonian thought was having generally, "Newtonianism, whether in its experimental or speculative aspect, barely figured in the thought of naturalists before 1730."

During the 1730s, however, Newtonian influences were reaching students of natural history and the number of more serious studies was growing (Hazard, 1946, Vol. 1, pp. 174-176). Mornet (1929, p. 86 ff.) says that toward 1750, Baconian and Newtonian ideas became "commonplace." Ritterbush may emphasize (1964, pp. 109-117) that "botanical analogy" was still the dominant vogue, and that in the work of Linnaeus we find still the search for a "divine plan for the creation" and the orthodoxy of graded function still enthroned; this is also the era, however, of the precise work of Réaumur, Trembley, Lyonet, the young Bonnet, and other careful experimentalists. Hazard tells us that geometry, geometric deductive reasoning, and Descartes had by then lost their supremacy to natural history, to Newtonianism, and to factualism. It would seem rather that there was neither a total dethronement of Cartesianism nor a triumph of new orthodoxies, but concepts roiling and clashing with a resultant melange of ideas. Even as late as 1787, several years after Abraham Trembley's death, his nephew Jean Trembley (1787, p. 23) still found it somehow necessary to criticize Cartesianism, contrasting his uncle's attitude with that of Descartes regarding "the advantages of philosophical doubt, that doubt which Descartes so extolled and of which he made such little use."

Before the French revolution, as Caullery (1933, p. 13) tells us, "a theological spirit" dominated the French universities. Thus prior to the nineteenth century, much of the scientific progress that took place came from the work of isolated individuals outside the universities who had other major occupations, such as law, medicine, and government. Guyénot (1941, pp. 189-190) describes the work on systematics and anatomy that was proceeding during this period in the universities and the medical schools; alongside it, but generally outside academia, studies were being carried out by the "curious," the observers and experimentalists who saw in organisms "something other than objects for collection," and who were excited by life processes, whether behavior, nutrition, reproduction, or metamorphosis. These "great connoisseurs" of nature were, says Guyénot, the "precursors of modern biology."

Growth of an International Scientific Community

Roger (1971, pp. 177-178) points out the decline of language barriers during this period as translations multiplied, learned men studied the living languages, and French became increasingly an international language. Periodicals were growing in number and influence. The "great observers" of nature in the mid-eighteenth century visited each other and corresponded and exchanged specimens. They translated each other's works and sought and recommended publishers for one another. They published reports which included news of colleagues' findings, and they corroborated and extended each other's observations. The following paragraphs afford a few examples.

When Bonnet discovered parthenogenesis in the course of studies suggested by Réaumur, the French savant invited the Genevans Trembley and Lyonet, both then resident in Holland, together with Bazin of Strassburg, to repeat Bonnet's experiments (Wheeler, 1926, p. 243). The first extended reports of Trembley's studies on the polyp were published by Réaumur (1742) in the preface to his sixth volume on the insects; some of Trembley's work on organisms other than the hydra is known to us only through Bonnet's publications (Baker, 1952, p. 113; Trembley, 1943, p. 286). In Trembley's preface to his own *Mémoires*, as we have mentioned, he presents recent work carried out by Lyonet on aphids. Bonnet and Trembley enjoyed a particularly close scientific friendship. In the *Mémoires* Trembley tells us that, as he came upon the discovery of regeneration in his polyps, he was influenced by Bonnet's discoveries on the reproduction of aphids to be open-minded. In the spring of 1743, Trembley wrote to Bonnet (translated in Dawson, 1983, p. 234):

I ardently wish that you would find some Polyps. I am convinced that you could be a great help to me if you had some, in order to complete my experiments . . . . I believe that there are wonderful discoveries to be made on all these Polyps; this is why I am extremely anxious for you to find some, for I am convinced that they could not fall into better hands.

In letters later that year Trembley offered to help Bonnet prepare the results of his own studies for publication (Dawson, 1983, p. 237 ff.).

In the *Mémoires* Trembley tells us not only of the assistance of Réaumur and Bonnet, but of other colleagues as well, such as J.N.S. Allamand, also in Holland as a tutor to the children of Leiden Professor 'sGravesande. Trembley relied on Allamand to repeat and verify his experiments on inverting hydras; Allamand carried them a step further, inverting animals that he had already inverted previously. Trembley and the Englishman John Needham were to have profound differences on the question of spontaneous generation, but in 1746 Trembley was scurrying about to have Needham's work not only translated, but also distributed in France (Needham, 1747; see Goeze, 1791, pp. xix-xx, Roger, 1971, p. 497, and Trembley, 1943, pp. 253, 285, 291-292).

Episodes alluded to in the Réaumur-Trembley correspondence (Trembley, 1943) illustrate many small triumphs of scientific internationalism over nationalist trade barriers and even acts of war. Beginning in April, 1744 the letters attest to interruptions in communication between French and British savants occasioned by the recurrent wars between their countries. To circumvent seizures of books, letters, and specimens carried abroad on French and British vessels, the scientists devised a system of double French and British addresses, which was effective only some of the time, and they relied heavily on such individuals as Trembley, welcome in both countries, to serve as go-betweens. As Trembley was about to depart from The Hague in June, 1745 for a stay in England, Réaumur wrote him: "All I ask of you regarding England is that you convey to Mr. Folkes for me as eloquently as you can my great esteem and affection for him. Though the English may hate the French, he seems to me no less worthy of being loved; men like him have nothing to do with the most implacable hatreds between nations" (Trembley, 1943, p. 239).

Folkes reciprocated such feelings, and together with Trembley and British officials, assisted at various times in the restitution to Réaumur of a variety of seized exotic birds, plants, insects, and even a young elephant en route from Senegal. (See Trembley, 1943, pp. 185, 214, 253, 296, 393-395, 397-399, 401-408, and 416.) As Roger (1971, p. 178) puts it, "Thus the Republic of Letters ignored frontiers, and the [warring] governments themselves seemed to grant it this privilege."

VII. POSSIBLE INFLUENCES ON TREMBLEY'S SCIENCE

Réaumur

In a letter to Réaumur of December 15, 1740 (see Trembley, 1943, p. 14) Trembley first describes the hydra to the French naturalist. He writes: "Since I am not learned in natural history, I am not aware of whether or not it [the hydra] is known." When he began his studies of the hydra in 1739, Trembley did not know of the much earlier discovery of the animal by Leeuwenhoek (1704), nor of the report by the "Anonymous Gentleman" on the hydra in the *Philosophical Transactions* of the Royal Society (1704). Guyénot (1943, p. X) views Trembley's knowledge of natural history at the time as "altogether elementary and superficial" and says that "Réaumur was his sole guide." Trembley himself expresses considerable diffidence about his limited knowledge of the subject in this and other letters of the period. Dawson (1983, p. 205) concludes, however, on the basis of her studies of the Réaumur-Trembley correspondence, that "Trembley's ingenuity in devising the experiments was clearly his own: Réaumur never suggested a particular experiment to Trembley."

His Mathematical Studies

Perhaps Trembley's mathematical studies as a university student served him better in the course of his investigations on the polyp than would have a focus on the less well-established zoology of the time. In that period the curriculum was still heavily classical, and the academic study of "small creatures" was not yet free of reliance on the ancient philosophers. Trembley, on the other hand, had been exposed to a far more rigorous and well-developed discipline when he studied under the mathematicians Cramer and Calandrini at the Geneva Academy of Calvin. That Trembley's thesis at the Academy dealt with the infinitesimal calculus may help illumine the precision and independent imaginativeness with which he approached problem-solving in his investigations of living organisms.

If one wished to apply the criteria developed by Thomas Kuhn in considering Trembley's contributions, one might observe with him that "Almost always the men who achieve these fundamental inventions of a new paradigm have been either very young or very new to the field whose paradigm they change . . . . for obviously these are the men who, being little committed by prior practice to the traditional rules of normal science, are particularly likely to see that those rules no longer define a playable game and to conceive another set that can replace them" (Kuhn, 1970, p. 90).

Dawson (1984, p. 45) sees a direct link between Trembley's mathematical studies and his conceptualization of his work with the polyps. Since "the discovery of infinitesimal calculus by Leibniz and Newton provided a striking mathematical justification for the idea of biological continuity," it may have been "no coincidence that Trembley moved from the exploration of mathematical to biological continuity." Baker (1952, p. 187) points out that Réaumur and Trembley both started as mathematicians. Jean Trembley (1787, pp. 7-8), Abraham Trembley's nephew who was himself a mathematician of note, assigns great importance to his uncle's study of mathematics, in particular his thesis on infinitesimal calculus, saying that "doubtless this study contributed more than everything else to inspiring in him the taste for that rigorous logic, that simple, lucid analysis, which shines in his works of natural history." Buscaglia (1985), in an illuminating analysis of Trembley's sectioning experiments, finds in them an "experimental logic" that is essentially mathematical in origin, based on Cartesian geometry, but with an added tendency "to go to the limits" that is reflective of Leibnizian calculus.

The Genevan Context

Leibnizian influences were certainly present in the lively intellectual mix that characterized the Geneva of Trembley's university years. That the "Empirical Newtonian tradition," channeled in part through Genevan connections with the

Leiden circle of Boerhaave and 'sGravesande among others, was definitely formative for Trembley and Bonnet is amply demonstrated by Dawson. (See especially Chapter III, "Geneva: The Cultural Matrix," in Dawson, 1983, pp. 58-94; also Dawson, 1985.)

Rudolph (1977) describes the remarkable intellectual flowering of Geneva during the eighteenth century, particularly in the life sciences. The Genevans of that era, he says, were "in the vanguard of physiological research." Among these men Rudolph lists Trembley, Bonnet, and Jean Senebier; he also comments on the early immunological work of Théodore Tronchin. Dawson refers to a Genevan "tradition in the natural sciences" which included Horace-Bénédict de Saussure, Pierre and François Huber, Jean-André de Luc, Nicolas de Saussure and Auguste de Candolle. In correspondence with the Genevan circle around Charles Bonnet were Albrecht von Haller, Lazarro Spallanzani, and Jean-Nicholas Sebastien Allamand (Dawson, 1983, pp. 5-7). In an article on Geneva in the *Encyclopédie* (see Rudolph, 1977, p. 50), d'Alembert remarks on the intellectual vitality both among natives and "famous foreigners." Perhaps the rich infusion of talent from descendants of Huguenot refugee families was at work here.

Religion

Mornet (1929, p. 85) tells us that before eighteenth century science could become truly experimental, it had first to reject its old dogma, idols, and goals of explaining the world, and to impose rigorous self-discipline upon itself. The attack on "scientific scholasticism" came not only from the philosophes, but also "from extremely pious people like the Abbé Pluche, Trembley, the Abbé Fromageot, president Rolland and twenty other teachers." If we emphasize Trembley's mathematical training and the diverse intellectual currents in his Genevan cultural background, if we depict the Trembley of the *Mémoires* as removed philosophically from many of his colleagues in his antipathy to much of the generalization and theory-building of the time, and in his refusal to bend scientific findings to the service of religious views, we in no way mean to suggest that he was some kind of modern sceptic. He shared with many of his more theoretically-inclined colleagues a common religious starting point, that is, a pietistic appreciation of the wonder and beauty of nature as reflections of the deity. Adhering to a non-sectarian form of Christianity, he also shared their arguments for God from the evidence of design and of order in nature.

Like the popular Abbé Pluche among others, Trembley also wrote of the value of the study of nature in developing morality and virtue in children. Trembley shared the religiously-inspired humility of other "modest" or "simplicist" naturalists of the period, though not the anti-intellectualism that many of them also expressed (Lovejoy, 1955, p. 66 ff.; Lovejoy, 1961, p. 7 ff.; Mornet, 1911, p. 140 ff.). Despite

Trembley's empiricism and the absence from the *Mémoires* of the theological interpolations found in much of the writing of the other "pious" naturalists, Trembley's work may still be viewed as part of the "insecto-theology" literature of the period, according to which the small creatures, with all their minute complexity and perfection, were emphasized as among the best examples of God's handiwork. This enchantment with the most minute elements of God's creation, implied by the subject of Trembley's *Mémoires* but never explicit in them, is expressed clearly by Trembley's fellow student of hydras, Henry Baker, in his *Employment for the Microscope* (1753). Baker says that all of God's creation is wonderful, but the tiniest "Specks of Life" by their "Minuteness" seem to embody "more Elegance and Workmanship (if the Term may be excused) in the Composition, more Beauty and Ornament in the Finishing" than he sees in the elephant, crocodile, and whale (H. Baker, 1753, p. 229). Such attitudes were in sharp contrast to "the contempt and neglect" which had been accorded the lower organisms for many centuries previous, as Wheeler (1926, p. 256) tells us.

In contrast to Bonnet and other theist naturalists of the time, Trembley appears to have had no difficulty in reconciling his discoveries with his religious beliefs. The *Mémoires* allude in only a few passages to God in terms of an infinite Creator of a complex and magnificent natural order. Trembley's later writings, however, show more fully his deeply religious orientation. According to Baker (1952, p. 41), from the start Trembley "regarded his scientific work as a religious exercise." Late in life after the death of a son, Trembley remarks in a letter that he finally has found some solace in turning to nature (Geisendorf, 1970, p. 282). Nature, god, science, faith appear to have melded meaningfully and comfortably for this antitheoretical, pragmatic observer.

In the course of our studies of Trembley, we often have wondered about the influence of his religious beliefs and attitudes on his science and have raised the subject with both scientific and lay colleagues. In this age of "secular science" it is probably no surprise that most with whom we spoke believed that a pious religious outlook was likely to interfere with conceptualizing, carrying out, and accepting scientific breakthroughs. Among scholars, Sigerist (1945, pp. 161-162) as an example, writes that science and medicine flourish in rationalistic, materialistic environments rather than where national philosophies are "mystical" or romantic. However, in the case of Trembley at least, we find ourselves believing that his religious views were not only a motivational, but also a scientifically liberating factor.

They were liberating because in the magnificent universe of Trembley's God all marvels are possible. By contrast, the philosophic or scientific rationalist is constrained by the necessity that a phenomenon appear "reasonable," that it fit with what is already accepted knowledge, and that it be subject to confirmation by inductive or deductive logic. As Carl Becker points out, the supposedly free-thinking philosophers of the eighteenth century were thus by no means free (Becker, 1967, pp. 102-103). It is instructive perhaps that Voltaire had great difficulties accepting Trembley's discoveries on the polyp. Trembley was undismayed by the disconcerting findings that did not seem to fit accepted understandings. On December 11, 1742 he wrote to Bonnet, "Your worm with two tails is admirable, but it does not surprise me, because nothing surprises me" (see Dawson, 1983, p. 164). Was it perhaps precisely because of his non-doctrinaire but profoundly religious world view that Trembley's mind and spirit were open to whatever observation and experiment might demonstrate to him?

The Relative Isolation of Sorgvliet

Another clue to the freshness of Trembley's approach may emerge from his sometimes disparaging remarks about "the prejudices of the schools." Several times in the *Mémoires* he points out that simple fishermen and young children recognize the truth for what it is in cases where academic theoreticians are oblivious to the facts before their noses. In such passages the author of our classically balanced, logical, and scholarly *Mémoires* seems surprisingly to verge on a kind of idyllic romanticized science in which Nature may most truly reveal herself to the simple, untutored soul. Nonetheless Trembley in no way really rejected the authority of the universities, the academies and societies, and the academicians. In the fourth Memoir, he quotes Professor Boerhaave of Leiden approvingly and at length, exclaiming: "What effort has this great man not expended in studying plants and animals!" (For a discussion of the possible influence of Boerhaave on Trembley, see Dawson, 1985.) Trembley's mentor, of course, is Réaumur, deeply involved with the Paris Academy. Through letters and later through his travels accompanying the young Duke of Richmond, Trembley communicated broadly with academics all across Europe. So, Trembley was hardly without academic interests and connections.

During those four intense years from 1740 to 1744 when Trembley was making and publishing his amazing discoveries, however, he worked in a setting remote from the universities and academies of Europe and from his Genevan circle. He was at that time serving as tutor to the two young sons of Count Bentinck of Holland and living in the Count's mansion at Sorgvliet near The Hague. In a letter from Trembley to his father (see Geisendorf, 1970, pp. 256-257), announcing that copies of the *Mémoires* were en route to him, Trembley writes:

You will see in them how I fill my leisure moments. ... I have found the means, by applying myself to the study of nature, of always having by me a thousand objects of recreation. All my glasses populated with little creatures are such good company with which to relax from more serious occupations.

Trembley sounds here very much like the country parsons and village doctors who communicated the findings of their Sunday nature jaunts to the Secretary of the Royal Society. Indeed, it was in the ditches of Sorgvliet that he first found the polyps, and during those "leisure moments" in the quiet of his study in the Bentinck mansion that he made his surprising discoveries. The vignettes at the head of each of the four Memoirs depict Trembley in the company of his two young charges. In the first three they are seen in various stages of collecting polyps or food for the polyps, and in the fourth they are shown in Trembley's study where an experiment on the inversion of a polyp is in progress. These vignettes are perhaps in instructive contrast to those in other works of the period which often depict groups of savants in discussion or observation while in the study of a king or in the meeting rooms of a scientific society.

Even during the years at Sorgvliet Trembley certainly was in contact both with great scientists and the virtuosi, as evidenced in his correspondence with Bonnet and Réaumur, and in such visits as that by the Duke of Richmond among others. But we find ourselves wondering if his physical and intellectual distance from the conflicting scientific schools of thought in the institutions of the city did not help him to approach his subject in a fresh, contemplative manner, inclining him to modes of thought relatively free of the crosscurrents of argumentation and pressures for recognition so prevalent in the culture of the university, scientific society, academy, and salon.

The scholarly enterprise in Trembley's time may have been simpler, but it appears to have been no more serene or idealistic than it is today. Between the giants Leibniz and Newton, for example, controversy had not too long since waged bitterly as to which of the two had precedence in formulating the calculus. A few years into the future was the notorious feud between Buffon and several leading scholars, the nastiness of which some commentators believe occasioned a virtual retirement from science for some years by one of Buffon's major targets, the great Réaumur himself (see Wheeler, 1926, p. 14 ff.). In the *Philosophical Transactions* of the Royal Society of the period, much is made of questions of priority of discovery regarding this finding or that. Some of this attitude appears even earlier in the *Transactions* of 1704 in the letter from the anonymous "Gentleman in the Country" who reports discovery of the polyps prior to Leeuwenhoek and states that he "was a little mortified to see . . . an account [Leeuwenhoek's] of a Creature which I thought that I had a sort of Propriety in, and of which I had made a Draught, with a design to present you and Mr. *C* with a rarity, which I believed no body had met with but myself." Trembley's relative isolation in Sorgvliet, then, may have afforded him some buffering from the public backbiting of the scientific societies and the salon as well as from the factious argumentation of the theoreticians of academia and may have made it easier for him to follow nature rather than fashion.

VIII. REACTIONS TO TREMBLEY'S DISCOVERIES

Though the study of nature always remained important to Trembley, and though he undertook some limited experimental work on and off throughout most of his life, yet as the Sorgvliet years wound down in the later 1740s, so did his systematic research. His preoccupations during his years of travel through Europe and his ventures into diplomacy and into scientific fields outside zoology may further incline some to regard Trembley as an amateur. It may be well to recall that many of the great minds of the period even among the professional scientists did not feel at all constrained to restrict themselves to pursuing a single discipline. It is also true, however, that despite his valuable discoveries of the mid-1760s, no scientific work of the scope and importance of the *Mémoires* followed from Trembley's pen after 1744. Perhaps as much as his antipathy to theoretical system building, this fact may also partially account for the manner in which Trembley's name faded from prominence during the nineteenth century.

Views of Several Scholars

Scholars have differed as to the nature of the immediate and long-term philosophical and scientific implications of Trembley's discoveries. Vartanian (1950, p. 253), for example, argues forcefully that the discovery of regeneration encouraged materialistic theories, and he certainly demonstrates a direct, incontrovertible influence at least upon that major exponent of mid-eighteenth century materialist philosophy, Julien Offray de La Mettrie. Ritterbush (1964, p. 127) counters that "the polyp converted no Englishman to materialism, since regeneration was still viewed as generally a vegetable property." He takes Vartanian to task for his error in stating that Leeuwenhoek called the polyp a plant, and then for constructing on Trembley's discovery of its animal nature his contention that "animal vitality came to be viewed as less complex."

There is ample evidence, however, for Bodemer's statement (1964, p. 21) that Trembley's "work profoundly agitated the imagination of his contemporaries and the polyp attained to great notoriety in the eighteenth century world." Certainly Trembley's discoveries created a stir in the universities, academies, salons, and studies of Europe. They also fed an already existing passion for studying lower forms. Archinard (1985) believes that Trembley's discoveries on the freshwater hydra were important in the evolution of the microscope. As microscopical studies on aquatic organisms became very fashionable in Britain in the wake of Trembley's investigations, naturalist John Ellis (who in 1755 published a study of "the corallines") commissioned John Cuff of London to construct the first aquatic microscope. Cuff was an outstanding instrument maker who had earlier created for

Henry Baker the first microscope having a true stage. The Cuff-Ellis aquatic microscope, which appeared in 1752, can be likened to a simple monocular dissecting microscope having one lens and a stage with a removable concave glass disc to hold the aquatic specimens.

If it is not yet clear just how directly Trembley's work influenced the development of the aquatic microscope, it is easy to document how Trembley's discoveries encouraged or led directly to a number of other important discoveries on a variety of small animals. Réaumur is a useful witness to both the public response to Trembley's work and the stimulus to new investigations. Himself responsible for helping to generate the excitement about the findings of the unknown tutor from Geneva, he writes, for example, (1742, Vol. 6, p. 1j ff.) that the discovery of regeneration is much the topic of discussion at court and in Paris, but that despite high interest the news has been greeted with considerable initial doubts. Réaumur considers the doubt a sign of healthy sophistication about such a finding which flies in the face of accepted ideas and "throws us into a new quandary regarding the nature of animals." He urges his correspondents, among them Jean-Étienne Guettard, Charles-René Girard de Villars, and Pierre Lyonet, to repeat Trembley's experiments. For other examples of work engendered by Trembley's findings see: pp. 38 and 41 of this Book; Goeze (1791) p. vi ff.; Kanaev (1969) pp. 8-10.

A Few Philosophical Repercussions of Trembley's Discoveries: Bonnet and his Circle

Trembley in his first Memoir quotes Réaumur's statement that even after seeing regeneration take place hundreds of times, he still is nearly incredulous each time he sees it. Many of the intellectual elite reacted feverishly to the mechanistic and materialistic implications that could be drawn from the discovery of regeneration in the polyp, a discovery "rich in speculative promise," as Vartanian (1963, p. 177) puts it. In a foppish style that seems worlds removed from the deliberative caution of his cousin, Charles Bonnet (see Trembley, 1943, p. 60) wrote to Professor Cramer of Geneva, under whom both he and Trembley had studied:

Now the animal studied by my cousin is thoroughly authenticated. Shall we attribute a soul to it, or none at all? . . . My great wish is only that my poor little creatures not be too much degraded . . . . I implore you, Sir, not to allow them to become simple machines. I will be inconsolable about it. Really, I will no longer observe them with as much pleasure. Good-by then to all industry, all skill, all kinds of intelligence.

Subsequently and somewhat more soberly, Bonnet (1744, pp. 479-480) brings these concerns to the Royal Society: "Where then does the Principle of Life reside in such Worms . . . . Are these Worms only mere Machines, or are they like more perfect Animals, a sort of Compound, the Springs of whose Motion are actuated by a kind of Soul?"

Dawson (1983, pp. 155-156, 167) shows us through her analysis of the correspondence of Réaumur, Trembley, and Bonnet, that Bonnet took a central role in the "metaphysical debate in Geneva over the implications of the discovery of the polyp" and that he "served as the transmitter of the Geneva interpretation to Trembley." Trembley's matter-of-fact responses to these metaphysical concerns were found to be annoyingly "laconic" in Bonnet's Genevan circle.

On such juicy philosophic issues of the time fed by Trembley's discoveries as questions of the materiality and divisibility of the soul, emboitement or preformation versus epigenesis, or the Ladder of Beings there is but little to be had from exploring Trembley's *Mémoires*. Later in life Trembley acknowledged a preformationist view in his only published venture into such speculative matters, but even then he did so acknowledging considerable qualms (Trembley, 1775, Vol. I, p. 326 ff. and Baker, 1952, p. 186). The *Mémoires* say nothing regarding preformation. There is a hint in the fourth Memoir, perhaps, of one of the considerations that may have moved Trembley into the preformationist camp. He derides "the hypothesis or rather prejudice" of spontaneous generation, and at the time that theory was interconnected with epigenetic concepts (see Baker, 1952, p. 185). Several writers on this point stress the views of Bonnet as having had a major influence on Trembley (see Mees, 1946, p. 150; Dawson, 1983, p. 155 ff.; Baker, 1952, p. 183 ff.). Also, the materialist connotations of epigenesis must have been troubling to the religious Trembley. It is ironic, therefore, that by the turn of the century, Trembley's discovery of regeneration had played an important part in bringing about the "demise" of preformationism which had "in its various forms dominated embryological thought for almost one hundred years before yielding to epigenesis" (Bodemer, 1964, p. 22).

The Chain of Being, and Trembley's Decision to Section the Hydra

Trembley's discovery of the regenerative powers of the polyp also was seized upon eagerly by many mid-eighteenth century theorists as exciting new evidence for the validity of the concept of the Chain or Ladder of Being. Prominent among them again was Charles Bonnet. To these theorists the polyp was the "zoophyte" predicated by the great Leibniz as the missing link that would be found between plants and animals (see Lovejoy, 1961, pp. 145, 194-195, p. 232 ff.). One will search in vain in Trembley's *Mémoires* and published letters, however, for allusions to this then-popular theory inherited from earlier centuries and carried to sometimes laughable extremes in the Age of Reason. In a recent discussion of the Chain of Being as it pertains to the *Mémoires*, Josephson (1985) finds that Trembley does not "seem to be much influenced by this concept." Trembley's one comment in the *Mémoires* on the zoophyte classification is to reject his own suggestions early during his investigations that the polyps might be so considered and urging instead that they be viewed as simple animals.

Since Trembley's time, it has often been assumed that the motivation leading him to his classical experiments on regeneration arose from his desire to place the polyp appropriately on the Ladder of Being. Most references recapitulate the statement that Trembley decided to section a hydra in order to prove to himself whether it was a plant or animal. As explained by Trembley in the *Mémoires*, the series of events leading to his decision to section the animal was more complex and interesting than that.

Trembley had early suspected that hydras were animals upon noticing that they contracted when their bodies were shaken, but he withheld judgment because of the possibility he had not yet eliminated "that they might be sensitive plants." After observing hydras carry out their step-by-step walking movements, however, Trembley wrote in the first Memoir (p. 6) that he was finally "persuaded . . . that they were animals." With that answer in hand, Trembley stopped studying hydra and turned his attention to other small creatures. He resumed his investigation of hydras only when he chanced to observe that the animals "gathered on the side of the jar facing daylight." From the time of his discovery of phototaxis in hydras, he "resolved not only to seek enlightenment on this matter, but also to try to delve deeply in general into the natural history of the polyp." His new attention was soon rewarded by an interesting observation (pp. 7-8) regarding the morphology of the animal:

It was not long before I noticed that not all the individuals of the species of polyps that I was observing have an equal number of arms or legs. I had reason to believe that there was nothing unnatural about these variations. Although I found no difficulty in accepting this difference among the individuals of a single species of animals, I nevertheless compared these arms at first with the branches and roots of plants, the number of which varies greatly among the individuals of the same species. At this point, I speculated anew that perhaps these organisms were plants, and fortunately, I did not reject this idea. I say fortunately because, although it was the less natural idea, it made me think of cutting up the polyps. I conjectured that if a polyp were cut in two and if each of the severed parts lived and became a complete polyp, it would be clear that these organisms were plants. Since I was much more inclined to think of them as animals, however, I did not set much store by this experiment; I expected to see these cleaved polyps die.

Thus the experiment of cutting the hydra was prompted by the morphological and developmental questions raised by the variations which Trembley saw in the numbers of tentacles in hydras of the same species rather than by any philosophical question regarding the polyp's place in the Chain of Being. If anything, once he had observed the varying numbers of tentacles, Trembley cut the polyp in order to further reinforce his conviction that the polyps were animals.

A Glimpse into Fashionable Society: Madame Geoffrin's Salon

If the reactions of Réaumur and Bonnet to Trembley's discoveries were of the order described above, what might be expected of individuals less versed in observation and experiment? From Mornet's analysis of library contents showing relatively few holdings of the *Mémoires* as compared with less rigorously factual writings on science (see p. 9 of this Introduction), one might wonder whether Trembley's "amazing findings" were more a voguish subject for discussion among the cognoscenti than his *Mémoires* were an object of serious study for them.

We have some curious testimony as to early reactions to Trembley's discoveries in the fashionable world from the irrepressible Madame Geoffrin, a great hostess of the philosophes in Paris. Characterized by Harcourt Brown (1940, pp. 225-226) as "One of the wisest and wittiest of the women of the eighteenth century," Madame Geoffrin had been asked for any new information on the subject of the polyps by her friend, Martin Folkes, President of the Royal Society of London. We hear echoes of the salon as she tells Folkes that Paris has been quite taken up with the business for some time, repeating gossip about who should have title to first discovery, "Mr. bonet" or a "Mr. tremblet" of Holland. She describes her discussion on the matter with Maupertuis, who did not seem particularly enthusiastic, viewing the "worms" in question as a species midway between plants and animals. She recounts her visit to Réaumur's study, still in search of information for Folkes, and reports that Réaumur will soon publish on the polyps. Most interestingly she asserts, with her distinctively creative grammar and spelling, that all this excitement is greater among "the ignorant" than among the academicians "for whom physics is the fashion." Likewise, in a letter of March 30, 1745 to Cadwallader Colden (see Ritterbush, 1964, p. 16), Peter Collinson writes: "The surprising phenomena of the polypus entertained the curious for a year or two but now the virtuosi of Europe are taken up in electrical experiments."

Thus we see that Trembley's relationships to some of the intellectual currents of his time were complex. Differences of opinion exist among scholars as to which players in the scientific scenario of the Enlightenment are to be deemed primarily Cartesian or Newtonian in outlook, as to whether the English or continental naturalists were the more empirical group, as to which philosophical trends were dominant, and so forth. Scholar Jacques Roger (1971, p. 161 ff.) uses the date 1745 to close one epoch in the development of the life sciences and open another. Trembley's work does indeed fit into Roger's characterization (1971, p. 451) of the earlier period as one that saw the triumph of a science based on observation as opposed to a priori mechanistic stipulations. We conclude that Trembley was certainly of his time and not out of step with it, but that the influences which shaped him were intriguingly diverse. We are inclined to think that Trembley reflects

the intellectual eclecticism in his Genevan education and background of which
Dawson (1983) writes. Trembley does not fit neatly into one philosophical or
practical niche. Although trained in mathematics, he favored observation and
experiment over any kind of abstract reasoning, whether inductive or deductive; he
most certainly valued knowledge of process and function over that of structure and
orderly classification. One of "the pious" in religious matters, he believed in the
study of nature as a means of glorifying God, but during his years of scientific
productivity he refused to mix religious questions, whether of animal soul or final
causes, into his science. He rejected the rationalist scholasticism found frequently
among the religious naturalists, signifying simply that in the Grand Design of the
Infinite Being all things were possible. He shared the humility of the "modest"
naturalists, but not their anti-intellectualism. In the quarrel of the Ancients versus
the Moderns, he paid due respect to the great Greek and Roman writers, but felt free
to gently mock their lack of scientific method. He believed in the ennobling quality
of scientific learning and saw the need for knowledge of science to be widely spread
and he wrote attractively for a broad audience, but never sacrificed accuracy for
style in the manner of a Buffon. Whether an amateur or not, he was a man of
exacting precision who produced solid, focused work, some of it revolutionary.

IX. IN THE SCIENTIFIC AFTERMATH

Studies of Regeneration

Trembley's work and studies by others pursuant to it brought significant
advances in the understanding of certain biological phenomena and of particular
organisms. The subject of regeneration is a useful example. Despite Réaumur's
work in 1712 on the regeneration of crayfish appendages, little attention had been
given to the phenomenon before Trembley's series of experiments on the polyp
carried out thirty years later. Thus, it was Trembley's experiments that stimulated a
spate of work on the subject of regeneration by Lyonet, Bonnet, de Jussieu,
Réaumur himself, and later Spallanzani (Guyénot, 1941, p. 201 ff.). As Bodemer
(1964, pp. 20-21) points out, "Thus, within twenty years of Trembley's experiments
on *Hydra*, regeneration was appreciated as a fundamental biological phenomenon
occurring in both invertebrate and vertebrate animals. This discovery tinted all
biological thought . . . ."

Early in the present century, we have only to look at the classic book titled
Regeneration by Nobel laureate Thomas Hunt Morgan (1901) to find an
appreciation by one of America's greatest biologists of Trembley's work on the
problem. In the first sentence of this book and in the first chapter in detail, Morgan
acknowledges Trembley's "celebrated experiments" as the first to bring attention to

the subject of regeneration, and as models, together with the work of Réaumur and Bonnet, furnishing the basis of all later work on the subject (Morgan, 1901, pp. 1-6). Interestingly, the state of knowledge of biology and the limited technology of that time apparently dissuaded Morgan from continuing his research on regeneration, and he began instead his work in genetics. There he gained his greatest fame in developing *Drosophila* (fruitfly) genetics, and in contributing greatly to the theory of the gene. Of this transition by Morgan in his research interests, John Tyler Bonner writes in the first paragraph of his stimulating monograph *Morphogenesis* (1952, p.3): "T.H. Morgan . . . is supposed to have said, in a lighter moment, that since he had been unable to solve the problem of regeneration . . . he had decided to try something easier such as the problem of heredity." Bonner then adds his own comment: "This sentiment, concerning the difficulty of the problem of regeneration and of development as a whole, is still as alive today as it was then."

Experimental Morphology; Physiology and Medicine

Some commentators credit Trembley as "the first to enter the realm of experimental morphology," but view that field as having subsequently remained virtually dormant for a century and a half until the work of Wilhelm Roux, followed by that of Spemann and others (Sirks and Zirkle, 1964, pp. 206, 246). Joseph Schiller (1974, p. 185 ff.) boldly dates "the birth of experimental biology" from the year 1740, asserting: "The birth certificate was signed by the Swiss naturalist Abraham Trembley who with his work on the freshwater hydra began the line of those naturalists who, starting from being passive observers of nature, became active experimenters. It was Trembley who initiated the trend, and not Spallanzani, as has been wrongly stated; Spallanzani came a quarter of a century later." Schiller describes as an immediate result of experimentation on the invertebrates the challenge to the presumed plan of nature based on plant versus animal classifications. Trembley's work, he says, also raised major problems regarding the individuality of the organism.

Rudolph traces a fairly direct, if sometimes quiescent lineage, from Trembley to Karl Ernst von Baer (see also Gasking, p.101 ff.). Von Baer (1792-1876), according to John Baker (1952, p. 47), was "generally regarded as the father of modern embryology," and von Baer in turn praised the work of Trembley as marking "the beginning of a new era in physiology" (Rudolph, 1977, p. 52), which was to have great influence on medicine. But Rudolph holds that the immediate consequences for medical research of Trembley's and Bonnet's work related to grafting were minimal, if not null. As Trembley turned to diplomacy, education, and government, and Bonnet away from research to theory and philosophy, the problem of "irritability" and the work of Haller commanded the greatest attention of the biological experimentalists of the second half of the eighteenth century (Rudolph, 1977, p. 55).

One may again pick up the thread near the close of the nineteenth century, a time according to Rudolph when "the work of Trembley and Bonnet was resumed with the same technique (which had retained all its freshness) and which was productive of the same kind of results." In this connection, Rudolph mentions such workers as George Wetzel and Eugen Korschelt. Also in this line of influence, says Rudolph, were such physiologists as Theodor Wilhem Englemann, Moritz Nussbaum, and Thomas Hunt Morgan (Rudolph, 1977, p. 55; see also Kanaev, 1969, pp. 385-388). Nussbaum's warm tributes in 1890 to Trembley and to the *Mémoires* (see Baker, 1952, p. 48) perhaps provide some validation of Rudolph's scientific genealogy:

In the year 1744 appeared Trembley's treatise, a masterpiece of precise presentation of carefully and prudently arranged observations, a classical model for a detailed biological investigation that undertakes to give in a single frame a picture of the whole life-history of a group of animals. Such a work, accompanied by the artistic illustrations of a Lyonet, will for ever [sic] remain, as regards form and content, a rich source of information for scientific research-workers, and will excite joyful admiration through the sincere modesty and scarcely surpassable clarity of the style.

Marine Studies: The Strange Case of Peysonnel, Marsigli, Réaumur, Trembley, and the "Coral Flowers"

Trembley's discoveries on the hydra and the bryozoan *Lophopus* stimulated new interest in marine as well as freshwater organisms. One specific immediate result was the reopening and resolution of the curious Peysonnel/Marsigli corals controversy.

In the preface to volume six of his *Mémoires*, Réaumur (1742, pp. 1xviij-1xxx) recounts, in an almost confessional manner, his curious role in the business of the "coral flowers." His remarks shed some interesting light on the state of the life sciences in the first half of the eighteenth century. Réaumur tells us that some fifteen years earlier, Jean-André Peysonnel, a physician from Marseilles, had informed him of his conviction that the corals were not plants as was commonly believed at the time, but animals. Count Marsigli, who was then deemed an authority on the matter, had categorically defined the "coral flowers" as plants. Réaumur, believing Peysonnel to be in error and not wishing the good doctor to be embarrassed by having his mistake made public, forwarded to the Paris Academy his own summary of Peysonnel's findings, without naming him, and with a critique of the idea that the corals were animals.

Jolted by Trembley's discoveries, in 1741 Réaumur began to be troubled by his earlier handling of Peysonnel's findings. He charged his trusted colleagues, de Jussieu and Guettard, with missions to the coasts of Normandy and Poitou to repeat, check, and extend Peysonnel's observations. In this manner, Peysonnel's work was vindicated and his findings published and gradually accepted by the scientific community. In his second Memoir, Trembley refers, with some satis-faction it seems, to the "fine observations" on corals by Réaumur, de Jussieu, and

Guettard. Réaumur paid direct tribute to Trembley for initiating the chain of events that led to the truth about the nature of corals and thus to Peysonnel's exoneration. (On the Peysonnel/Marsigli matter, see also Trembley, 1943, pp. 116-117; Guyénot, 1941, p. 89.)

Describing how he has himself been stimulated by Trembley's work to make new observations on marine polyps, Réaumur (1742, pp. lxxix-lxxx) predicted that naturalists would seek out additional species of marine polyps and study differences in their morphology, feeding, growth, and reproduction: "At last, a part of natural history that is so interesting and new and that has been sketched only roughly, will be studied thoroughly as it deserves."

X. OUR SEARCH FOR MATERIALS RELATED TO ABRAHAM TREMBLEY

Our search began in 1954 when H.M.L., then a new Ph.D. in biochemistry and the first student of the late W.F. Loomis, began to investigate the hydra. In Loomis's laboratory, requisite reading for any initiate to research on hydras was the biography of Abraham Trembley published by John R. Baker only a few years previously (1952). After our first reading of the biography, we could not help noting two curious similarities between the situations of Trembley undertaking research on hydra in the mid-eighteenth century, and of Loomis at the outset of the renaissance of hydra study in the mid-twentieth century. Each man conducted his research on an estate in relative isolation from academia, and while acting as mentor to one or two students. Trembley was employed as a tutor by the wealthy Count Bentinck at his Sorgvliet estate. In Loomis's case, he was a wealthy scientist running his own independent laboratory in suburban Connecticut, and was mentor to H.M.L. [Fuller accounts of the scientific contributions emanating from four generations of scientists in the Loomis family can be found in Alvarez (1980) and Lenhoff (1983b).]

A few years after leaving the Loomis laboratory, we obtained a volume of the 1744 Leiden edition of the *Mémoires* and began our translation in bits and pieces. A major stimulus for our finally completing the translation came via the efforts of Trembley's great-great-grandson, Maurice (1874-1942). Maurice Trembley spent more than 40 years collecting and annotating the letters of Abraham Trembley and his correspondents. His invaluable collection of letters between Abraham Trembley and Réaumur was published one year following his death (Trembley, 1943). We were introduced to the work of Maurice by his son, the late Jean-Gustave Trembley, and his family. They literally opened their home and archives to us so that we might review and photograph letters and documents related to Abraham Trembley, including such treasures as the unpublished correspondence between Abraham Trembley and Martin Folkes, then President of the Royal Society. For some details of our search for documents related to Trembley, see Lenhoff (1980).

With our search ended and the *Mémoires* translated, our link to the Trembley family was renewed once again in December of 1984 at an international symposium held in Geneva to commemorate the two hundredth anniversary of Trembley's death. At that time, the city, under the auspices of the Société de Physique et d'Histoire naturelle de Genève and the University of California, Irvine, dedicated a commemorative plaque at the Trembley School. The proceedings of the symposium are published as Volume 38, number 3, of the *Archives des Sciences* (1985).

XI. NOTES ON OUR APPROACH AS TRANSLATORS AND EDITORS

Our first concern in translating the *Mémoires* was to provide a version that would be easy for the contemporary reader of English to understand. Happily, Trembley's organization of the *Mémoires* is generally clear, logical, straightforward, and uncluttered; his language is simple, direct, and supple; and he is little given to the dramatic apostrophes and circumlocutions often encountered in similar writings on science during this period.

Wheeler (1926) provides an excellent commentary on the scientific style of the time in the introduction to his edition of Réaumur's *The Natural History of Ants*. It is not surprising that a few of the less attractive elements of Réaumur's style, as described by Wheeler, do also occasionally appear in Trembley's writing. These include "the heaping up of relative clauses, the looseness of syntactical construction, the avoidance of concisely periodic sentences and a preference for indirect or negative statements." Wheeler describes these as "the expression of a calmer, more leisurely and more refined reaction to the social environment" than that to which the modern reader is accustomed. Though "desire for completeness of description" of which Wheeler speaks seems to lead Trembley occasionally to pile clause upon clause in a paragraph consisting of one long sentence, or to extend an unbroken paragraph for several pages, Trembley is not given to the "slow movement and diffuseness" that Wheeler and others (see also Mornet, 1911, pp. 11, 202) attribute to Réaumur. Nor is Trembley's prose "very restrained, tenuous and almost atonic, like a minuet performed on an old harpsichord." There is some harpsichord in the writings of Trembley, but no minuet; to a remarkable degree the sound seems to us modern, bright, and fresh.

We have taken occasional liberties with Trembley's sentence structure and paragraphing when we believed that the "heaping up" was troublesome and that we could enhance the clarity of the narrative by so doing, and we have rather reluctantly left out even most of those harpsichord echoes. Measurements are rendered into their closest modern equivalents. A *ligne* is rendered as 2.26 millimeters despite the anachronism entailed. (The metric system was not introduced until near the end of the eighteenth century.) Since *pouce* (1.0657 of an English inch) and inch are so close, however, we retained the nonmetric language in this instance.

We have tried to avoid most such anachronisms while striving for clarity for today's reader, however. Some of our decisions on usage may surprise certain of our readers, and it may be helpful for them to understand how we arrived at them. Our translation of the term *insectes* as *small creatures* may be taken as a case in point. The French word was sometimes rendered into the English of the time as *insect*. It was also sometimes rendered as *little creature* (see Baker, 1743, p. 3 ff., for example), a term that we believe conjures up a more appropriate image for the modern reader. An amazing range of animals was covered by the word *insecte* in mid-eighteenth century usage. According to Réaumur (1734, Vol. 1, p. 58), "A crocodile would be a prodigious insect, but I would have no difficulty in calling it one." In addition to reptiles, Réaumur's *insectes* could include arachnids, worms, "polyps," molluscs, crustaceans, and, in fact, any animal that was neither fish, bird, nor quadruped.

Seeing that the word *fecundity*, like the word *insect*, was used quite broadly at the time of Trembley, we translate the term variously according to the context in which it is used, as pertaining to fertilization, reproductive capacity, or reproduction. In regard to such usage and the vocabulary of the *Mémoires*, we were fortunate to have for purposes of comparison original copies of the 1743 book by Henry Baker titled *An Attempt towards a Natural History of the Polype* and a French version by Demours of Baker's book published a year later as *Essai sur l'Histoire naturelle du Polype, Insecte.*

Another usage that some readers may find questionable is our rendering of *corps organisé*, as *organism*. Relevant English documents of the period that we examined, such as the French communications translated in the *Philosophical Transactions* and Baker's volume on the polyp, do not use the literal rendering *organized body*, often encountered in modern translations of the eighteenth century French. One finds in these mid-eighteenth century English versions *minute body*, *small body*, or *body*, terms which because they may be used also at other times by these writers to designate inanimate objects, do not immediately convey the sense of a living organism. Numerous examples can be found in Volume 42 and 43 of the *Philosophical Transactions* in translations from the writings of Réaumur, Bonnet, and Trembley, and in articles by Folkes and Needham. [See Volume 42, Number 467, pp. xiii (293), xv (295); Number 469, pp. 423, 433; Number 470, p. 468; Number 471, p. 639; Volume 43, Number 474, pp. 171, 173, 174, 181.] Thus we made the admittedly debatable decision to translate *corps organisé*, as *organism*. (See, for example, pp. 7 and 10 of the translated *Mémoires*.)

We found little need to break into the text of the *Mémoires* with aids for the reader other than to insert current page references in brackets where Trembley speaks of points "elsewhere," "later," or "previously" in the *Mémoires*. Where Trembley is inconsistent in italicizing titles or denoting quoted material, we have simply standardized the usage without indicating these small discrepancies in the original text. Errata are handled with the reader's ease in mind. We found no more than a dozen, and these we surmise to be mostly overlooked typesetting errors. They are listed and explained on page 60. In the text we give the correct usage as based on context and scientific accuracy, indicating the errata by asterisks. These errors lie mostly in the labeling of figures and in confusion of the terms *inverted* and *everted*.

We have translated everything in the *Mémoires* including Trembley's quotations taken from other works, except for specific book titles. Thus, the few Latin lines from Ovid that Trembley incorporates and Trembley's quotations from Boerhaave and Réaumur are in our words.

Our handling of calendar dates in the *Mémoires* needs a few words of explanation. Britain remained on the Julian Calendar till after 1750 whereas most European countries had long since adopted the Gregorian or "New Style" Calendar. Thus at the time the *Mémoires* were published the British calendar was eleven days earlier than the continental one. In the 1744 translation of Henry Baker's volume on the polyp from English into French, Demours (p. 8) changes Baker's English dates into dates "according to the new style." In our translation, however, we have chosen not to shift to the English calendar; we simply retain the "new style" dates used by Trembley.

Our secondary aim, after that of access for today's reader, was to try to capture whatever of the grace and charm of the original *Mémoires* we could, both physically and textually. Mornet tells us (1911, pp. 199, 202) that it was only during the eighteenth century that writers on science began to be anxious about pleasing their readers stylistically. Prior to that era science was "not an affair of literature," it was mostly expounded in Latin, and it was not intended for a general public. Biographer Baker (1952, p. 41) speaks of Trembley's style as "generally limpid in its clarity." We wished to tamper with it as little as possible. We opted for language reflecting usage of the era when it seemed to us not to interfere with the reader's understanding of the material. Thus, we used the term *powder jar* for *poudrier* despite Wheeler's (1926, pp. 230-231) cogent argument for translating it as *beaker*. We were mindful, however, of those scientists (see pp. 2-3) who tell us that they value Trembley's writings not so much for their historical interest as for the still useful information they contain, and that understanding served to curb proclivities toward the quaint or towards favoring style over precise meaning that we might have been tempted to indulge. We have relied on Dr. Baker (1952) for modern identification of most of the organisms cited in the *Mémoires*, and have inserted his designation in brackets alongside Trembley's at various critical points in the translation. A summary of Dr. Baker's identifications is presented in the Glossary (p. 48).

Bodenheimer (1958, p. 29) cautions that it "is indispensable to read the publications under study as a contemporary would have read them when they were first published. This is by no means an easy task." We believe the task he would set us to be not merely difficult, but probably impossible. We agree instead with Mornet's remarks (1929, p. 165) on the matter: "We are never certain of understanding men of genius as they would have wished us to understand them, and we are certain of not understanding them as their contemporaries did." Though in this introduction we have offered the reader some of our reflections on the scientific zeitgeist, on the philosophical, religious, and educational milieu from which Trembley produced the *Mémoires*, and on the reactions to him of his contemporaries, we do so believing these help us but to guess at the eighteenth century ambience.

We hope in our translation that we have allowed Trembley to speak to the modern reader in a reasonable facsimile of his own voice, for we can imagine no better way for that reader to enjoy and benefit from the translated *Mémoires* than to follow this gentle scholar who writes:

The facts that I must report are too extraordinary to ask that anyone take my word for them. I shall explain as clearly as possible every consideration that guided me, and all the precautions I took to avoid self-deception. Insofar as I am able, I shall bring the reader into my study, have him follow my observations, and demonstrate before his eyes the methods I used to make them. He himself will be witness to my results.

XII. GLOSSARY

In Table 3 we list in one column our translation of the wording Trembley uses in his *Mémoires* to describe various organisms; in the second column we match to them the names with which Dr. John Baker (1952) identifies those organisms. In the third column we add some brief descriptions to aid the nonbiologist.

Table 3. Glossary of names of organisms

Names Used by Trembley	Baker's Identifications	Description
Beetle, to which Trembley tried to feed the polyps	*Gyrinus natator*	A whirligig
Bivalve shell, creature that lives in a	An ostracod	Small, free-swimming crustacean
Caddis worm	Aquatic larva of a caddis fly	Makes and lives within a temporary casing composed of gravel and other materials; Plate 10, Fig. 1
Crane fly	See Midge (of the Tipulidea)	
Duckweed	*Lemna*	See Plate 1, Figs. 1 and 7
Fleas, water	*Daphnia pulex*	See Plate 6, Fig. 11
Fleas, water, with branched horns	Same as above	
Gnat, transparent larva of	Aquatic larva of the gnat *Chaoborus*	
Horsetail plants	*Myriophylum*	A water-milfoil
Lice	*Kerona polyporum*	An aquatic protozoan; Plate 7, Figs. 8 and 9
Midge, red larva of	Larva of midge *Chironomus*	
Millepede	*Stylaria lacustris*	A freshwater oligocheate segmented worm; Plate 6, Fig. 1
Millepede, barbed	Same as above	
Polyp, first species	*Hydra viridissima*	See Plate 1, Fig. 1
second species	*Hydra vulgaris*[1]	See Plate 1, Fig. 2
third species	*Hydra oligactis*	See Plate 1, Fig. 3
with long arms	*Hydra oligactis*	
marine	squid, octopus, cuttlefish	
tufted	*Lophopus crystallinus*	A colonial bryozoan (of the Ectoprocta); Plate 10, Figs. 8 and 9
Slugs, flat black	*Polycelis*	A flatworm (triclad); Plate 7, Fig. 9
Spiders, of red color	Red hydrachnids	Small brightly colored aquatic mites
Worm	Usually *Tubifex*	A slender aquatic worm; Plate 7, Fig. 2
Worm, flat, white (with red viscera)	*Glossiphonia*	A lightly colored flat leech; Plate 7, Fig. 7. Often feeds on larvae of the midge *Chironomus* (see below)
Worm, thick red	Larva of the midge, *Chironomus*	See Plate 7, Fig. 8, and Plate 11, Fig. 12

1. Alternate name *Hydra attenuata.*

XIII. REFERENCES

Adams, George. 1746. Of the fresh water polipe, with arms in form of horns. Pages 138-166 *in* George Adams, *Micrographia Illustrata, or, the Knowledge of the Microscope Explain'd*. Published by the author, London.

Alvarez, L.W. 1980. Alfred Lee Loomis. Biographical memories. *Natl. Acad. Sci. U.S.A.* 51:309-341.

Anonymous. 1704. Two letters from a gentleman in the country, relating to Mr. Leuwenhoeck's letter in Transaction No. 283. *Phil. Trans.* 23:1494-1501.

Archinard, Margarida. 1985. Abraham Trembley's influence on the development of the aquatic microscope. Pages 335-344 *in* H.M. Lenhoff and P. Tardent (eds.) *From Trembley's Polyps to New Directions in Research on Hydra. Archives des Sciences*, Vol. 38, Fasc. 3.

Baker, Henry. 1743. *An Attempt towards a Natural History of the Polype*. R. Dodsley, London.

Baker, Henry. 1744. *Essai sur l'histoire naturelle du polype, insecte*. (Transl. from English by M.P. Demours.) Durand, Paris.

Baker, Henry. 1753. *Employment for the Microscope*. R. Dodsley, London.

Baker, John R. 1952. *Abraham Trembley of Geneva: Scientist and Philosopher 1710-1784*. Edward Arnold & Co., London.

Barzansky, B., and Lenhoff, H.M. 1974. On the chemical composition and developmental role of the mesoglea of hydra. *Am. Zool.* 14:575-581.

Becker, Carl L. 1967. *The Heavenly City of the Eighteenth-Century Philosophers*. Yale University Press, New Haven. [Copyrighted in 1932.]

Bentinck, the Honourable William, Esq; F.R.S. 1744. Abstract of part of a letter from the Honourable William Bentinck, Esq; F.R.S. to Martin Folkes, Esq; Pr. R.S. *Phil. Trans.* 42:ii (= 282).

Blanquet, R., and Lenhoff, H.M. 1966. A disulfide-linked collagenous protein of nematocyst capsules. *Science* 154:152-153.

Boas, Marie. 1958. *History of Science*, Publication Number 13. American Historical Association, Washington, D.C.

Bodemer, Charles W. 1964. Regeneration and the decline of preformationism in eighteenth century embryology. *Bull. of the Hist. of Med.* 38:20-31.

Bodemer, Charles W. 1967. Overtures to behavioral science: eighteenth and nineteenth century ideas relating light and animal behavior. *Episteme* 1:135-152.

Bodenheimer, Friedrich Simon. 1958. *The History of Biology: An Introduction*. Wm. Dawson and Sons Ltd., London.

Bonner, John Tyler. 1952. *Morphogenesis: An Essay on Development*. Princeton University Press, Princeton, New Jersey.

Bonnet, Charles. 1744. An abstract of some new observations upon insects communicated in a letter to Sir Hans Sloane, Bar[t] late President of the Royal Society, etc. (Transl. from French by P.H.Z. Esq; F.R.S.) *Phil. Trans.* 42:458-488.

Brown, Harcourt. 1940. Madame Geoffrin and Martin Folkes: six new letters. *Mod. Lang. Qtrly.* 1:215-241.

Buscaglia, Marino. 1985. The rhetoric of proof and persuasion utilized by Abraham Trembley. Pages 305-319 *in* H.M. Lenhoff and P. Tardent (eds.) *From Trembley's Polyps to New Directions in Research on Hydra. Archives des Sciences*, Vol. 38, Fasc. 3.

Campbell, Richard D. 1983. Identifying hydra species. Pages 19-28 *in* H.M. Lenhoff (ed.) *Hydra: Research Methods*. Plenum Press, New York.

Caullery, Maurice. 1933. *La science francaise depuis le XVIIe siecle*. Libraire Armand Colin, Paris.

Dawson, Virginia P. 1983. *The Animal Machine and the Problem of the Polyp in the Letters of Bonnet, Trembley and Réaumur*. Ph.D. Thesis. Case Western Reserve University, Cleveland, Ohio.

Dawson, Virginia P. 1984. Trembley, Bonnet and Réaumur and the issue of biological continuity. *Studies in Eighteenth-Cent. Cult.* 13:43-63.

Dawson, Virginia P. 1985. Trembley's experiment of turning the polyp inside out, and the influence of Dutch science. Pages 321-334 *in* H.M. Lenhoff and P. Tardent (eds.) *From Trembley's Polyps to New Directions in Research on Hydra. Archives des Sciences,* Vol. 38, Fasc. 3.

Ewer, R.F. 1949. Famous animals—1. Hydra. Pages 100-111 *in* L.M. Johnson and Michael Abercrombie (eds.) *New Biology.* Penguin Books, London.

Folkes, Martin to Abraham Trembley. November 30, 1743. Archives of the Jean-Gustave Trembley Family, Geneva.

Folkes, M. 1744. Some account of the insect called the fresh-water polypus, before-mentioned in these Transactions. *Phil. Trans.* 42:422-436.

Freedman, Arthur, ed. 1966. *Collected Works of Oliver Goldsmith.* Vols. 1 and 5. The Clarendon Press, Oxford.

Gasking, Elizabeth. 1970. *The Rise of Experimental Biology.* Random House, New York.

Geisendorf, Paul-F. 1970. *Les Trembley de Genève de 1552 à 1846.* Alexandre Jullien, Genève.

Goeze, J.A.E. 1791. see Trembley, A. 1791.

Goldsmith, Oliver. See Freedman, Arthur.

Goss, Richard J. 1969. *Principles of Regeneration.* Academic Press, New York.

Gronovius, J.F. 1744. Extract of a letter from J.F. Gronovius, M.D. at Leyden, November 1742, to Peter Collinson, F.R.S. concerning a water insect, which, being cut into several pieces, becomes so many perfect animals. *Phil. Trans.* 42:218-220.

Guyénot, Emile. 1941. *Les sciences de la vie aux XVIIe et XVIIIe siècles.* Editions Albin Michel, Paris.

Guyénot, Emile, 1943. See Introduction in Trembley, Maurice. 1943.

Hahn, Roger. 1971. *The Anatomy of a Scientific Institution: The Paris Academy of Sciences, 1666-1803.* University of California Press, Berkeley, California.

Hazard, Paul. 1946. *La pensée européenne au XVIIIeme siecle, de Montesquieu à* Lessing. Ancienne Librairie Furne, Boivin & Cie, Paris.

Hessinger, D.H., and Lenhoff, H.M. 1976. Membrane structure and function: Mechanism of hemolysis induced by nematocyst venom: Roles of phospholipase A and direct lytic factor. *Arch. Biochem. Biophys.* 174:603-613.

Josephson, Robert. 1985. Old and new perspectives on the behavior of hydra. Pages 347-358 *in* H.M. Lenhoff and P. Tardent (eds.) *From Trembley's Polyps to New Directions in Research on Hydra. Archives des Sciences,* Vol. 38, Fasc. 3.

Kanaev, I.I., see Trembley, A. 1937.

Kanaev, I.I. 1969. *Hydra: Essays on the Biology of Fresh Water Polyps.* Edited by Howard M. Lenhoff. (Transl. from Russian by Earl J. Burrows and Howard M. Lenhoff.) Published in limited edition by Howard M. Lenhoff. [Originally published by Soviet Academy of Sciences, Moscow, 1952.]

Krinsky, N., and Lenhoff, H.M. 1965. Some carotenoids in hydra. *Comp. Biochem. and Physiol.* 16: 189-198.

Kuhn, Thomas S. 1970. *The Structure of Scientific Revolutions.* University of Chicago Press, Chicago, Ill. [Copyrighted in 1962.]

Leeuwenhoek, Antony van. 1704. Concerning green weeds growing in water, and some animalcula found about them. *Phil. Trans.* 23:1304-1311.

Lenhoff, Howard M. 1959. Migration of C^{14}-labeled cnidoblasts. *Exp. Cell Res.* 17: 570-573.

Lenhoff, Howard M. 1961. Digestion of ingested protein by hydra as studied by radioautography and fractionation by differential solubilities. *Exp. Cell Res.* 23:335-353.

Lenhoff, Howard M. 1965. Cellular segregation and heterocytic dominance in hydra. *Science* 148:1105-1107.

Lenhoff, Howard M. 1969. *p*H profile of a peptide receptor. *Comp. Biochem. Physiol.* 28:571-586.

Lenhoff, Howard M. 1978. The hydra as a biological control agent, pages 58-61, *in Mosquito Control Research.* Edited by R.E. Fontaine. University of California Extension, Berkeley.

Lenhoff, Howard M. 1980. Our Link with the Trembleys—Abraham (1710-1784), Maurice (1874-1942), and Jean-Gustave (1903-1977). Pages xvi-xxiv *in* P. Tardent and R. Tardent (eds.) *Developmental and Cellular Biology of Coelenterates.* Elsevier/North Holland Biomedical Press, Amsterdam.

Lenhoff, Howard M. 1981. Biology and physical chemistry of feeding response of hydra. Pages 475-497 *in* R. Cagan, (ed.) *Biochemistry of Taste and Olfaction.* Academic Press, New York.

Lenhoff, Howard M. (ed.) 1983*a. Hydra: Research Methods.* Plenum Press, New York.

Lenhoff, Howard M. 1983*b*. William Farnsworth Loomis (1914-1983)—In Memorium. Pages ix-xxiv *in* H.M. Lenhoff (ed.) *Hydra: Research Methods.* Plenum Press, New York.

Lenhoff, Howard M. and Brown, R. 1970. Mass culture of Hydra: Improved method and application to other invertebrates. *Laboratory Animals* 4:139-154.

Lenhoff, Howard M. and Lenhoff, Sylvia G. 1984. Tissue grafting in animals: Its discovery in 1742 by Abraham Trembley as he experimented with hydra. *Biol. Bull.* 166:1-10.

Lenhoff, Howard M. and Loomis, W.F. (eds.) 1961. *The Biology of Hydra and of Some Other Coelenterates: 1961.* University of Miami Press, Coral Gables, Florida.

Lenhoff, Howard M. and Tardent, Pierre (eds.) 1985. *From Trembley's Polyps to New Directions in Research on Hydra. Archives des Sciences,* Vol.38, Fasc. 3.

Lenhoff, Howard M., Kline, Edward S., and Hurley, Robert E. 1957. A hydroxyproline-rich, intracellular, collagen-like protein of *Hydra* nematocysts. *Biochim. et Biophys. Acta* 26:204-205.

Loomis, W.F. 1954. Environmental factors controlling growth in hydra. *J. Exptl. Zool.* 126:223-234.

Lovejoy, Arthur O. 1955. *Essays in the History of Ideas.* George Braziller, Inc., New York.

Lovejoy, Arthur O. 1961. *The Great Chain of Being, A Study of the History of an Idea.* Harvard University Press, Cambridge, Massachusetts. [Lectures delivered at Harvard University in 1933; copyrighted in 1936.]

Mees, Charles E.K. 1946. *The Path of Science.* John Wiley and Sons, Inc., New York.

Miall, Louis Compton. 1912. *The Early Naturalists, Their Lives and Work (1530-1789).* Macmillan, London.

Morgan, Thomas Hunt. 1901. *Regeneration.* The Macmillan Company, New York.

Mornet, Daniel. 1911. *Les sciences de la nature en france au XVIIIᵉ siecle, un chapitre de l'histoire des idées.* Librairie Armand Colin, Paris.

Mornet, Daniel. 1929. *La pensée francaise au XVIIIᵉ siecle.* Librairie Armand Colin, Paris.

Muscatine, L., and Lenhoff, H.M. 1963. Symbiosis: On the role of algae symbiotic with hydra. *Science* 142:956-958.

Needham, T. 1747. *Nouvelles découvertes faites avec le microscope.* Luzac, Leide.

Novak, Patricia and Lenhoff, Howard M. 1981. Induction of budding in a nonbudding mutant of *H. viridis* as a result of grafts using wild-type tissue with varying compositions of interstitial and nerve cells. *J. Exptl. Zool.* 217:234-250.

Réaumur, René Antoine Ferchault de. 1734-1742. *Mémoires pour servir à l'histoire des insectes.* Imprimerie Royale, Paris, 6 vols.

Ritterbush, Philip C. 1964. *Overtures to Biology: The Speculations of Eighteenth-Century Naturalists.* Yale Univ. Press, New Haven, Connecticut.

Roger, Jacques. 1971. *Les sciences de la vie dans la pensée francaise du XVIIIᵉ siecle.* Armand Colin, Paris.

Rudolph, Gerhard. 1977. Les débuts de la transplantation expérimentale: considérations de Charles Bonnet (1720-1793) sur la "greffe animale." *Gesnerus* 34:50-68.

Rushforth, N.B., Burnett, A.L., and Maynard, R. 1963. Behavior in hydra: Contraction responses of *Hydra pirardi* to mechanical and light stimuli. *Science* 139:760-761.

Rutherford, C.L., Hessinger, D., and Lenhoff, H.M. 1983. Culture of sexually differentiated hydra. Pages 71-78 *in* H.M. Lenhoff, (ed.) *Hydra: Research Methods.* Plenum Press, New York.

Schiller, Joseph. 1974. Queries, answers and unsolved problems in eighteenth century biology. *Hist. Science* 12:184-199.

Sigerist, Henry E. 1945. *Civilization and Disease*. Cornell Univ. Press, Ithaca, New York.

Sirks, M.J. and Zirkle, Conway. 1964. *The Evolution of Biology*. The Ronald Press, New York.

Stimson, Dorothy. 1948. *Scientists and Amateurs: A History of the Royal Society*. Henry Schuman, New York.

Trembley, Abraham. 1744a. Observations and experiments upon the fresh-water polypus. (Transl. from French by P.H.Z.) *Phil. Trans.* 42:iii-xi (= 283-291).

Trembley, Abraham. 1744b. *Mémoires, pour servir à l'histoire d'un genre de polypes d'eau douce, à bras en forme de cornes*. Verbeek, Leyde.

Trembley, Abraham. 1744c. *Mémoires pour servir à l'histoire d'un genre de polypes d'eau douce, à bras en forme de cornes*. Durand, Paris. 2 vols.

Trembley, Abraham. 1744d. Letter to Jean Trembley, December 1, 1744. Archives of the Jean-Gustave Trembley Family, Geneva.

Trembley, Abraham. 1775. *Instructions d'un père à ses enfans, sur la nature et sur la religion*. Chapuis, Genève. 2 vols.

Trembley, Abraham. 1791. *Des Herrn Trembley Abhandlungen zur Geschichte einer Polypenart des süssen Wassers mit hörnerförmigen Armen*. (Transl. from French by J.A.E. Goeze.) Reussners, Quedlinburg. [First published in 1775.]

Trembley, Abraham. 1937. *Mémoires, pour servir à l'histoire d'un genre de polypes d'eau douce, à bras en forme de cornes*. (Transl. into Russian by I.I. Kanaev.), State Biological and Medical Lit. Press, U.S.S.R.

Trembley, Jean. (Anon.) 1787. *Mémoire historique sur la vie et les écrits de Monsieur Abraham Trembley*. Fauce, Neuchatel.

Trembley, Maurice. 1943. *Correspondance inédite entre Réaumur et Abraham Trembley*. (Introduction par Emile Guyénot.) Georg, Genève.

Vartanian, Aram. 1950. Trembley's polyp, La Mettrie, and eighteenth century French materialism. *J. Hist. Ideas* 11:259-286.

Vartanian, Aram. 1963. *Diderot and Descartes, A Study of Scientific Naturalism in the Enlightenment*. Princeton University Press, Princeton, New Jersey.

Wheeler, William Morton, (ed.) 1926. *The Natural History of Ants, From An Unpublished Manuscript in the Archives of the Academy of Sciences of Paris, by René Antoine Ferchault de Réaumur*. Alfred A. Knopf, New York.

INDEX

EDITORS' OUTLINE OF CONTENTS

MEMOIR IV

XVI. A GUIDE FOR READING
THE TRANSLATED *MÉMOIRES*

As pointed out by Baker (1952, p. 174), the accuracy of Trembley's observations "is perhaps the most striking feature of Trembley's work." Nonetheless Trembley did make some factual errors, and Baker (1952, p. 175) lists eight that Trembley made in regard to those experiments on the hydra reported in the *Mémoires*. Most of those were trivial. The few major ones were either the result of the low power and resolution of Trembley's magnifying glass and microscope, or of Trembley's extreme caution in not drawing conclusions unless the evidence was complete.

Three of the errors, however, may confuse those readers who are not fully familiar with the biology of the hydra. Hence, we list those errors, as described by Baker, and make some suggestions that may facilitate the reading of the translated *Mémoires*. The errors are:

a. "He did not distinguish clearly between the nematocysts [stinging capsules] of the ectoderm and the carotene granules of the endoderm."

b. "He did not clearly express the existence of the two *layers* in the body-wall that are nowadays called ectoderm and endoderm, though he knew that the two *surfaces* were different in structure."

c. "He considered that after 'reversal' or snipping into small fragments, the outer surface of the body could take on the function of the inner, and *vice versa*."

The first and second errors described above should not be confusing so long as the reader is aware of them. These distinctions (a and b) were not recognized until about 100 years later when better histological methods had been developed. In fact, the first error regarding the nematocysts and the carotene granules did not alter the veracity of other conclusions that Trembley reached while studying the granules, i.e. regarding the materials holding the granules together (protoplasm), and regarding the location and uptake of pigment in the granules of the inner surface. Nonetheless, the reader should be aware of Trembley's thinking while reading pp. 26-27 and 30-33 of Memoir I and pp. 80-81 of Memoir II.

The third error, on the other hand, is confusing and disrupts the flow of Memoir IV. Thus, we suggest that at first the reader may prefer to skip over pp. 152-154, which deal with the regeneration of a complete polyp from a thin strip by "inflation" of that strip, and pp. 163-165, which deal with many variations of the inversion experiments.

XVII. ERRATA

We found only nine errata in the *Mémoires*. In these instances, we substitute the correct word, lettering or number, and mark each with an asterisk. The first case appears to be an error in using the number 19 instead of the number 9; the larger number does not make sense in the context in which it is used. The second through the fifth errors deal with references to the labeling of Fig. 8 of Plate 10. The last four errata deal with the confusing section on inverting hydras, where even an expert in this area must pay close attention to the figures and the text so as not to confuse the terms inverted, everted, non-inverted, and non-everted. We list the errata and the corrections in the following table.

Table 4. Errors and corrections

Page	Error	Correction
106	19	9
129	*i k l b i m*	*i k l b i m*
	i o g	i g b
	a g	a g
	a g	a b
167	non-everted	everted
	everted	non-everted
168	non-everted	everted
171	inner	outer

BOOK II

A TRANSLATION FROM THE FRENCH

OF

ABRAHAM TREMBLEY'S

*MÉMOIRES, POUR SERVIR À L'HISTOIRE
D'UN GENRE DE POLYPES D'EAU DOUCE,
À BRAS EN FORME DE CORNES*

MEMOIRS

CONCERNING THE

NATURAL HISTORY

OF A TYPE OF

FRESHWATER POLYP

WITH ARMS SHAPED LIKE HORNS

by A. TREMBLEY, of the Royal Society

Leiden
Jean and Herman Verbeek
M. DCC. XLIV.

Translated and Edited by
SYLVIA G. LENHOFF and HOWARD M. LENHOFF
Department of Developmental and Cell Biology, University of California, Irvine

PREFACE.

OMME il y a divers Animaux qui peuvent être placés dans la Claſſe des Polypes, j'ai cru qu'il étoit néceſſaire, pour déſigner ceux dont il eſt queſtion dans ces *Mémoires*, d'ajouter au nom general de Polype, l'indication de quelques-uns de leurs caractéres. Je leur ai donc donné le nom de *Polype d'eau douce*, pour les diſtinguer des Polypes de mer. Mais, cette diſtinction n'étant pas ſuffiſante, parce qu'on trouve dans les eaux douces pluſieurs genres d'Animaux, qui paroiſſent devoir être rangés dans la Claſſe des Polypes, j'ai cherché dans les Polypes, dont je dois parler, un caractére qui pût ſervir à indiquer même leur genre, & à le diſtinguer des autres genres de Polypes d'eau douce. Ils n'en ont point de plus facile à remarquer, que celui que préſente la forme de leurs bras: & c'eſt pour cela, que je les ai appellés *Polypes d'eau douce à bras en forme de cornes.*

LES Obſervations, que renferment ces *Mémoires*, ſont le fruit des Recherches, que j'ai faites ſur ces Ani-

maux,

PREFACE.

N ORDER to distinguish between the particular polyps discussed in these *Memoirs* and various other animals which could be included in this class, I thought it necessary to add some description of their specific traits to the generic name of polyp. Thus, I called them *freshwater polyps* to distinguish them from polyps of the sea. This distinction proved inadequate, however, as fresh water harbors many kinds of animals that apparently ought to be classified as polyps. In the polyps I am to discuss, I sought a trait that not only would clearly characterize the species but also would distinguish it from other freshwater varieties. Since their most obvious feature is the shape of their arms, I named them *freshwater polyps with arms shaped like horns.*

THE observations contained in my *Memoirs* are the fruit of three and a half years of research which I carried out on these animals. Having ascertained that the polyps possess the remarkable capacity to multiply as a result of being cut into sections, I became curious to delve deeper into their natural history. Thus, I set out to study them with care and diligence. I

PREFACE

did not at first imagine that these animals would keep me busy for so long a time. During the first two years of studying them, I was swept along, as it were, from one observation to another with barely the time to make notes in my journal. For this reason, I was unable to begin preparing at an earlier time the *Memoirs* that I now present to the public.

Even had I enjoyed the leisure to draft these *Memoirs*, it is easy to understand why I would have been very reluctant to do so while in the midst of making interesting observations. These observations could not help but furnish me with new material, and even influence to some extent the plan I would need to follow in writing the *Memoirs*. Moreover, since at the start of my observations the polyps were altogether unknown to me, I needed to concentrate on even the most minute particulars in order not to overlook any which might be important. If, therefore, I had written during those early stages, I would have run the risk of elaborating on a great number of insignificant facts that need but be touched upon in the process of disclosing more important matters. Only after enlarging our knowledge of the subject at hand are we in a position to consider such details in their true perspective. Furthermore, in order to verify certain findings, I was obliged to wait from one year to the next that I might make identical observations under precisely identical circumstances.

However rough the account of my observations on the polyps might have been, I would have hastened to make it public if I had not had other means of reporting the most interesting findings. It seemed to me from the start of my observations that knowledge of the remarkable properties of the polyps could bring pleasure to the inquisitive and contribute something to the progress of natural history. Therefore, I set myself the task of communicating my discoveries as I made them. Insofar as I could, I gave polyps to those wishing to repeat my experiments, and I explained my procedures to them.

Consequently, in a short time the polyps became rather widely known, and observers in a number of locations were in a position to verify some of my experiments, as did Mr. Baker in England last summer. Shortly thereafter he published the journal of his observations to which he added a summary of the information on the polyps already printed both in the preface to the sixth volume of the *Memoirs on Insects* by Mr. Réaumur and in numbers 467 and 469 of the *Philosophical Transactions*.

It would have been impossible for me to explain my observations clearly without the help of a great number of illustrations. I was as fortunate in this

PREFACE

respect as I was in my discoveries on the polyps. Without flattering the talented artist to whose kindness I owe the figures accompanying this work, I can attest that no one else could have equaled him in executing the drawings needed to clarify my descriptions. I had but to put the objects before Mr. Lyonet for him to see everything that would have been difficult to point out to others. Not only is he a skillful draftsman, but also a keen and experienced observer.

In addition to drawing the figures that accompany this work, Mr. Lyonet engraved the last eight plates. I know of no better way to praise the beauty of his engraving than to refer to the plates themselves. To see them is to admire them.

It may seem surprising that Mr. Lyonet did not engrave the first five, but my explanation will be even more surprising. The drawings for these five plates were ready for engraving in May of 1743. At that time, Mr. Lyonet did not yet know how to engrave, nor had he ever seen the process. Nevertheless, by the following month he was already an excellent engraver. As difficult as it may be to believe what I have said, it is nonetheless a fact supported by several witnesses. In May, 1743, chancing to meet Mr. Wandelaar, who is one of the most accomplished engravers in Holland, Mr. Lyonet showed him a number of his drawings depicting small creatures that he had observed. Struck by the beauty of these drawings and recognizing in them great skill and aptitude, Mr. Wanderlaar was convinced at once that Mr. Lyonet could as readily excel in engraving as he did in drawing. He urged him to give it a try. Thus it happened that in the middle of the next month (June 1743) Mr. Lyonet arrived at the home of Mr. Wandelaar in Leiden. It was then, for the first time in his life, that he handled the etching needle and the graver. His first work, an engraving of a dragonfly, could easily be taken for that of a master. This success persuaded Mr. Lyonet to assemble the tools needed for engraving. As soon as he had them, he immediately engraved three butterflies which provided further proof of his extraordinary talent. Finally he undertook the engraving of the polyps. He began with the sixth plate which he executed in September. The seventh was completed in October, the eighth, ninth, and tenth in November and December, and the last three in the following January and part of February.

Considering the large number of figures found in these plates and the delicacy of the engraving, one would be inclined to think that Mr. Lyonet had been busy with them during the entire period which I have just described,

PREFACE

that is from September 1743 until the middle of the following February. He spent, however, but a small part of that time on this work, just a part of the leisure that his various pursuits allowed him.

Based on this fact and on all that I have said concerning Mr. Lyonet's progress in such a difficult art, one may well conclude that his engravings are as extraordinary in their own way as are the animals which they depict.

Mr. Lyonet now is beginning to engrave the small creatures that he himself has studied, the drawings of which are admired by all who see them. How highly desirable it would be should his pursuits permit him to publish a complete work. It would fully merit the interest of all those who love the sciences and the arts, and would be extremely valuable in accelerating the progress of natural history. No one is better able to make such a contribution than a skillful observer who is at the same time an excellent draftsman and engraver.

I am very pleased that Mr. Lyonet has permitted me to report a discovery he has made on a subject which has been attracting the well-deserved attention of naturalists for some years. Findings about the reproduction of various kinds of terrestrial aphids are presented on page 556 in the sixth volume, thirteenth *Memoir* of the work of Mr. Réaumur. In it there is mention of the very interesting observation by Mr. Lyonet and Mr. Bonnet, that viviparous aphids also produced oblong objects which appeared to be eggs. In relating this fact, Mr. Réaumur questions whether these are really eggs. Mr. Lyonet's idea, from the moment he saw them, that they were indeed eggs has now been verified, at least as regards the particular species of aphids discussed by Mr. Réaumur starting on page 334 in volume thirteen, ninth *Memoir* of his work on insects, and illustrated in figures 5, 6, 7, 8, 9 and 10 of the 28th plate.

In April, 1743, Mr. Lyonet noticed ants gathering on the bark of an oak tree and became curious to know what was attracting them there. He found some small oblong, brown objects, which he at once surmised to be the eggs of aphids. He brought them home, kept them with care, and at the end of a fortnight, he did indeed see aphids emerging from them. These he recognized as the species described by Mr. Réaumur in the reference cited above. Mr. Lyonet showed me one of the aphids as it was coming out of the egg. A number of eggs remained on one of the oaks where Mr. Lyonet had found them. The aphids which hatched from them fed upon this oak and multiplied greatly. It is noteworthy, however, that from April until the end of

PREFACE

September, not a single egg was ever found on the bark of the oak. To the contrary, one often found mothers giving birth to young. As the weather turned colder, neither aphids nor eggs could be detected. Nevertheless, there were eggs on that oak tree, for at the start of spring, just as in the preceding year, their existence was revealed by the presence of ants clustering around them. The young aphids hatched at the beginning of May. It is to be noted that when the eggs were found, no young at all were seen, and that when young were found, the eggs were no longer seen. These observations, which Mr. Lyonet intends to repeat, would lead us to the following conclusions: this kind of oak aphid is viviparous in summer, and oviparous towards the end of autumn. The last generation of the year produces eggs and not young. During the winter, it is these eggs which preserve the generation that is to give rise to young aphids in the spring.

The first five plates, which Mr. Lyonet did not engrave, are the work of Mr. van der Schley, a skillful disciple of the famous Mr. Bernard Picart. Although Mr. van der Schley began engraving small creatures but a short time ago, he already has attained a degree of excellence that demonstrates what may be expected from him in this field. At a time when increasing attention is given to the study of small creatures, it is extremely fortunate that those studying the natural history of these animals have such a talented worker to help them.

Mr. van der Schley, using the drawings of Mr. Pronk, also engraved the vignettes found at the beginning of each of these *Memoirs*. These copperplate engravings very accurately depict scenes of Sorgvliet, the country house of Count Bentinck. It is there that I found the polyps, and there that I made most of the observations discussed in these *Memoirs*. My two pupils often joined me there in fishing for polyps and in hunting for various other small creatures. In their company I often witnessed how even children can begin to appreciate the pleasures of contemplating nature. To a child, nature presents a pageant which at first entertains him but then spurs his curiosity, instructs him, enchants him, moves him, and accustoms his spirit to delight in all that is most beautiful.

The vignette at the beginning of the first *Memoir* relates to the information in that *Memoir* and in the following one about fishing for polyps. It shows three people by the bank of a fishpond in Sorgvliet, two of whom are searching for polyps. One is looking for them on the aquatic plants that he holds in his hand; the other facing him is trying to find them on the plants in a jar filled with water. I found polyps very frequently in the fishpond, one end of which is shown in this vignette. A part of the greenhouse of the Sorgvliet orangery is visible from the spot where the three figures stand.

The subject of the second vignette is taken from the *Memoir* which it introduces. On the bank of the pond located in the middle of the menagerie of Sorgvliet, a man is using the implement described in the second *Memoir* on page 61 to fish for water fleas. He is shown just at the moment that he is drawing the hoopnet used for catching them out of the water. Two instruments for collecting water fleas are propped against the first two trees on the left.

The third vignette deals with collecting worms, as described on page 62 of the second *Memoir*. A man standing on the bank of a fishpond is dipping an instrument for collecting worms into the water. This tool, described on the page just cited, is like the one propped against the second tree to the left. Alongside the man is the jar into which he puts the worms after drawing them from the pond. Two young people sitting close by are amusing themselves by examining the worms already in the jar. This vignette and the one heading the first *Memoir* each show a different end of the same pond.

In the vignette preceding the fourth *Memoir*, a man is standing before a table absorbed in carrying out the most difficult procedure discussed in this *Memoir* (p. 154). He is turning a polyp inside out. The polyp lies on the edge of his open left hand, and he is inverting it with a hog bristle held in his right hand. The sun entering the room through the window at the man's right shines on the polyp, illuminating it for the experimenter and for the child examining it through a magnifying glass. In the two large jars on the sill of the wide open window, the artist has attempted to show groups of polyps like the one drawn lifesize in Plate 9. Alongside the jar in the middle of the sill stands a glass vessel enclosed in the cardboard sleeve which is discussed on page 37 at the end of the first *Memoir*. The chevron-shaped opening on one side of the sleeve is distinctly visible. It is at this opening that the polyps gather. In this illustration, the chevron faces the room. To perform the experiment proving that polyps have a proclivity for light, however, it is necessary to turn the opening toward broad daylight. Upon this same window sill I long kept my very first jar of polyps, the ones discussed on pages 5 and 7 in the first *Memoir*.

CONTENTS

OF THE

MEMOIRS.

FIRST MEMOIR.

SECOND MEMOIR.

THIRD MEMOIR.

FOURTH MEMOIR.

C. Pronk del. ad vv. 1744.   J. v. Schley sculp.

MEMOIRES
POUR L'HISTOIRE
DES POLYPES.

✶❁❁❁❁❁❁❁❁❁❁❁❁❁❁❁❁❁❁❁❁❁❁❁❁✶

PREMIER MÉMOIRE.

Où l'on décrit les Polypes, leur Forme, leurs Mouve-
mens, & une partie de ce qu'on a pu découvrir
fur leur Structure.

ES Faits auffi finguliers, auffi contrai-
res aux idées généralement reçues fur
la nature des Animaux, que le font
ceux que m'a fait voir l'Infecte dont je
vais donner l'Hiftoire, demandent, pour
être admis, les preuves les plus évidentes. Il eft ar-
rivé plus d'une fois, que la précipitation, & l'amour

A   du

C. Pronk del. ad viv. 1744.   J. v. Schley sculp.

MEMOIRS

CONCERNING THE NATURAL HISTORY

OF THE POLYPS.

* * * * * * * * * * *

FIRST MEMOIR.

*Wherein are described the polyps, their form,
their movements, and some discoveries
about their structure.*

he little creature whose natural history I am about to present has revealed facts to me which are so unusual and so contrary to the ideas generally held on the nature of animals, that to accept them demands the clearest of proofs. More than once, haste and a predilection for the fantastic have

led naturalists into error and have concealed from them what they otherwise could have recognized easily. It is not enough to say, therefore, that one has seen such and such a thing. This amounts to saying nothing unless at the same time the observer indicates how it was seen, and unless he puts his readers in a position to evaluate the manner in which the reported facts were observed.

As for me, I need to adhere to this rule as strictly, and even more strictly, than others. The facts that I am going to report are too extraordinary to ask that anyone take my word for them. I shall explain as clearly as possible the details of how I was led to discover them and all the precautions I took to avoid self-deception. Insofar as I am able, I shall bring the reader into my study, have him follow my observations, and demonstrate before his eyes the methods I used to make them. He himself will be witness to my results.

Still, I would not have been satisfied to begin this treatise had I been obliged to rely solely on my own observations to establish the truth of the extraordinary natural phenomena these *Memoirs* encompass. As I sensed from the outset, such findings need more than one eyewitness to be believed. If at first I could scarcely believe what I was seeing with my own eyes, I had to think that others would find it all the more difficult to believe what only I had seen. Thus I did not neglect any opportunity to have others see all that I had seen.

In this respect I was as fortunate as one could be. The individuals who were kind enough to evaluate my observations, and those who independently repeated them unquestionably may be counted among the finest experts. One of these well-known names in itself carries the full weight of authority.

Although I had excellent witnesses available in this country, I would have been subject to just criticism had I been slow to rely on the testimony of the great naturalist who is the pride of France and of his century. Had I not already had the good fortune to be corresponding with him, I would have had ample reason to ask this favor of him so that he could evaluate my observations and see for himself. Obviously I speak of Mr. Réaumur. Everyone who has any taste for natural history is acquainted with the delights of his books; every naturalist whose research has had some success feels obligated and honored to acknowledge the debt he owes him for the inspiration of his books and for the great lessons they contain.

Thus, on discovering the first of the exceptional traits of the little creatures in question, one of my first concerns was to inform Mr. Réaumur, and I continued to communicate my subsequent findings to him. In addition, I sent some of these animals to him. Using them and others that he found in abundance around Paris, he tried most of the experiments I had performed. Mr. Réaumur gave these animals the name of *polyps*. In the course of this work I will show that it would have been impossible to choose a more fitting name.

To my knowledge all those who have repeated my experiments have obtained the same results as I did. The experiments which are of special concern here and which have been repeated most often by others consist of cutting the little animals *transversely and lengthwise*, either in half or into a number of parts. The result of these experiments is that each part develops into a perfect animal through a very obvious reproduction of whatever is needed to make a complete polyp.

The public has already learned of the results achieved by Mr. Réaumur's experiments in particular. Here is how he puts it in the preface to the sixth volume, page 55 of his *Mémoires sur l'Histoire des Insectes*. "I must admit that the first time I saw two polyps gradually take form from the one which I had cut in half, I could scarcely believe my eyes. It is a phenomenon I simply cannot get used to seeing, even after seeing it again and again hundreds and hundreds of times." Mr. Réaumur cut some polyps into a number of parts, and each part became a whole polyp. He also informed the public that no sooner had this amazing reproduction been recognized in the polyps, than it was detected by him and other observers in various species of worms as well. In two years' time, this phenomenon has become widely known, so that these facts which at first seemed beyond belief, by now have been proven true of a variety of animals differing not only in species, but even in genus. From all indications, this property will yet be discovered in a great number of other creatures.

The preface by Mr. Réaumur I have just cited is not the only published corroboration of many of the strange facts of natural history which I am going to discuss. Another quite valuable account is in Number 469, Article 6, of the *Philosophical Transactions*. Since Mr. Folkes, President of the Royal Society of London, wished to see with his own eyes what he had heard about the polyps, in February, 1743 I sent some to him which arrived safely. He immediately cut them up in different ways and in a short time he saw the pieces of polyp become complete polyps. In the article from the *Philosophical Transactions* just cited, Mr. Folkes reports to the Royal Society the results of experiments he performed during the first fifteen days that he observed the polyps. To read this report is to be thoroughly informed regarding a good part of the polyp's natural history and to be convinced of the reality of most of the strange facts of that history.

These animals were not hitherto entirely unknown. They are mentioned in the *Philosophical Transactions* of 1703 (No. 283, Art. IV, and No. 288, Art. I). The observations on these little creatures made by Leeuwenhoek and by an anonymous Englishman are recorded there. There is much consistency between the observations of these two gentlemen. Both noticed one of the most remarkable characteristics of the polyps, that is, their natural mode of multiplying. They were struck by it and certainly would not have failed to study it further had they possessed a substantial number of polyps. Leeuwenhoek, however, was able to find only a few, and the anonymous Englishman but a single polyp. They described the external appearance of the polyps and some of their movements.

Mr. Bernard de Jussieu has been familiar with polyps for a long time; he even had drawings made of some, as Mr. Réaumur tells us (Preface to Volume VI of his *Memoirs on Insects*, p. 54). Beyond that, I know that some other people saw polyps before I did, and presumably many more inquisitive individuals would have chanced upon them in the course of investigating plants or aquatic animals.

None of these observers noticed that remarkable process of reproduction which takes place in the various parts of a polyp which have been severed. Because of its nature, that finding was to be not the fruit of long patience and great wisdom, but a gift of chance. It is to such a happy chance that I owe this discovery which I made, not only without forethought, but without my ever having had in my entire life any idea even slightly related to it.

Every unusual discovery naturally arouses curiosity about the way in which it was made. This I could appreciate from the questions addressed to me by a considerable number of people. First they asked me how I happened to think of cutting up the polyps, and how I came to observe this reproduction by which many parts of the same polyp each becomes a complete animal. Accordingly, I decided to begin the account of my observations on the polyp by answering this question.

Since my first summer in 1740 at Sorgvliet, the country house of Count Bentinck, situated a quarter of a league from The Hague, I have found polyps there. Having noticed various small animals on the plants that I had taken from a ditch, I put some of these plants into a large jar filled with water, placed it on the inside sill of a window, and then set about examining the creatures that it contained. Soon I discovered a great many of them, all quite common indeed, but most of them unfamiliar to me. The novel spectacle presented me by these little animals excited my curiosity. As I scanned this jar teeming with creatures, I noticed a polyp fastened to the stem of an aquatic plant. At first I paid it little heed, for I was following the livelier little creatures which naturally attracted my attention more than an immobile object. The casual observer, especially one completely unfamiliar with such physically similar animals as marine polyps, could scarcely avoid taking the freshwater polyp for a plant. I have said that the polyp I had noticed was motionless. The point is not that it was unable to move, but at that time I knew nothing about whether it could move or not.

Before proceeding further, I must, for the sake of clarity, describe here the general structure of these animals. The body a b (Plate 1, Fig. 1) is quite slender. From one of its extremities a, the horns a c project; they serve as arms and legs, and are even more slender than the body. I call extremity a anterior, because the head of the polyp is there; to the opposite extremity b, I give the name posterior. When discussing the parts formed by sectioning through a polyp transversely, I shall call the area around the head the first part, the next portion the second part, and so forth.

At the start, I did not encounter all three species of polyps with arms shaped like horns which I shall mention in this work. I first became acquainted with the smallest species. The polyps of this species are a rather pretty green color. These are the particular polyps that I am now discussing. There were many of them in the large jar which I described as well as in another vessel in which I had also placed some aquatic plants.

The first times that I observed these little organisms, I took them to be parasites growing on the other plants. They were in the position shown in Figure 1 of the first Plate.

The shape of these polyps, their green color, and their immobility gave one the idea that they were plants. This same initial impression was evoked in many people who saw the polyps for the first time when the creatures were in their most usual position. Some said that the polyps were bits of grass; others compared them to the tufts on dandelion seeds.

The first movement that I noticed in the polyps was in their arms, which bent and twisted slowly in all directions. Thinking that the polyps were plants, I could scarcely imagine that this movement was their own. Yet those slender threads projecting from one of their extremities did appear to move by themselves and not in response to the agitation of the water. I suspected, however, that the other creatures swimming about in the same jar might agitate the water enough to produce the perception of movement in the arms of the polyps. The more I followed the movement of these arms, however, the more it seemed that it had to come from an internal cause, and not from an impetus external to the polyps.

One day I jogged ever so slightly the vessel holding the polyps in order to see how the ensuing movement of the water would affect their arms. I was completely unprepared for the result. I expected to see their arms and even their bodies merely shaken and dragged along with the motion of the water. Instead I saw the polyps contract so suddenly and so forcefully that their bodies looked like mere particles of green matter and their arms disappeared from sight altogether. I was caught by surprise. This surprise served to excite my curiosity and make me doubly attentive. Without stopping, I ran my eyes, with the help of a magnifying glass, back and forth over many of the polyps I had caused to contract, and I soon noticed some beginning to extend. Their arms again became visible, and little by little these polyps resumed their original shape.

This contraction and all the movements I saw the polyps make as they extended once again roused sharply in my mind the image of an animal. I likened them at first to snails and other creatures that contract and extend.

One may find it amazing, perhaps, that I did not conclude definitely that polyps were animals. I was still influenced, I admit, by their shape and their green color. I thought it not impossible that they might be sensitive plants and I found their contraction and extension no more extraordinary than the response of such sensitive plants to touch. This idea, then, kept me doubting, and I preferred not to come to any decision until new observations would resolve the issue.

At the end of several days I found a number of polyps fastened to the sides of the vessel in a location where I had seen none before, and where there certainly had been no polyps at all at the start. I soon learned how they had come there. While I was watching, several of them advanced along the walls of the vessel. I shall describe their manner of walking elsewhere [p. 19] and be content to say here that they take steps in much the same way as do inchworms and various aquatic insects, that is, by successively attaching their anterior and posterior ends, the posterior after drawing close to the anterior, and the anterior after moving a distance away from the posterior.

The sight of this step-by-step movement of the polyps finally persuaded me that they were animals. Once convinced, I stopped observing them for I had found what I was seeking. Until then I had intended no more than to learn whether they were animals. Almost the entire month of September, 1740 passed without my giving them the least attention. During that time I was preoccupied with other creatures which I had been observing for a long time. Toward the end of that month, however, the polyps once again attracted my attention, this time so strongly that I have not stopped observing them since.

Recall now that my glass vessel had been placed on the inside sill of a window. One day I noticed a great number of polyps gathered on the side of the jar facing the daylight. I was at first curious to learn whether their clustering was merely fortuitous, or whether it resulted from a marked propensity of the polyps for the best-lit area of the jar. To satisfy my curiosity, I turned the jar halfway around so that the large cluster of polyps was on the least bright side of the jar, and only a few polyps were on the side facing the light. Then it was a question of seeing whether the mass of polyps would pass from the poorly lit side to the best-lit side. The day after turning the jar I found that the poorly lit side on which I had left many polyps was almost entirely devoid of them. The polyps were dispersed in the jar on their way to the best-lit side; for the next day I found a number of them already there, and after a few days, as many polyps were on this side as had previously been on the other before I had turned the jar halfway round. Once more I turned the jar and in so doing I repeated the same experiment and I witnessed the same result. After seeing the same thing a number of times, I became convinced that the polyps had a distinct propensity for the best-lit area of the jar. I did not venture to decide whether this propensity was directly related to the light, or whether some other factor attracted them to the best-lit side. At a later point [p. 37] I shall recount at greater length the observations I made on this subject which at that time struck me as worth my attention. From that moment on I resolved not only to seek enlightenment on this matter, but also to try to examine the overall natural history of the polyps.

It was not long before I noticed that not all the individuals of the species of polyps that I was observing have an equal number of arms or legs. I had reason to believe that there was nothing unnatural about these variations. Although I found no difficulty in accepting this difference among the individuals of a single species of animals, I nevertheless compared these arms at first with the branches and roots of plants, the number of which varies greatly among the individuals of the same species. At this point, I speculated anew that perhaps these organisms were plants, and fortunately I did not reject this idea. I say fortunately because, although it was the less natural idea, it made me think of cutting up the polyps. I conjectured that if a polyp were cut in two and if each of the severed parts lived and became a complete polyp, it would be clear that these organisms were plants. Since I was much more inclined to think of them as animals, however, I did not set much store by this experiment; I expected to see these cleaved polyps die.

On November 25, 1740 I sectioned a polyp for the first time. I put the two parts into a shallow glass which contained water to the height of only about nine to eleven millimeters. In this way I could observe the parts of the polyp easily with a rather strong magnifying glass.

I shall describe elsewhere the precautions I took in performing my experiments on these cut polyps as well as the manner in which I set about cutting them. For now, suffice it to say that I cut the polyp transversely a little closer to the anterior than to the posterior end. Thus the first part was a little shorter than the second.

The instant I cut the polyp the two parts contracted so that at first they looked like no more than two small granules of green matter at the bottom of the glass into which I put them. As I have said, the first polyps I cut were green in color. The two parts extended the same day that I separated them. They were quite easy to distinguish from one another because the first had its anterior end bedecked with those fine threads which serve as the polyp's arms and legs, whereas the second had none at all.

The act of extending itself was not the only sign of life that the first part gave on the day it was separated from the other, for I saw it move its arms. On the following day, the first time that I came to observe it, I found it had changed its place, and shortly afterward, I saw it take a step. The second part remained extended as on the preceding day and in the same spot. I shook the glass a little to see if it was still living. When this movement caused it to contract, I concluded that it was alive. A short time later it extended anew. On the days that followed I saw the same thing occur.

Still I considered the movement of the two parts of a single polyp merely as signs of a feeble remnant of life, especially as regards the second part. Since I was presuming that the polyp was an animal, I expected its head to be located on the anterior part, as indeed it is. It seemed natural enough to me that the half composed of the head and a portion of the body could still live. I thought that the operation which I had performed had only mutilated the head part without essentially disrupting its animal economy. I compared this first part to a lizard which has lost its tail and which does not die from losing it. Indeed, again supposing that the polyp was an animal, I assumed that the second part was only a kind of tail without the organs vital to the life of an animal. I did not think that it could survive for long separated from the rest of the body. Who would have imagined that it would grow back a head! I was observing this second half to find out how long it would retain the remnants of life; I had not the least expectation of being a spectator to this marvelous kind of reproduction.

I observed these parts through a magnifying glass several times each day. On the morning of December fourth, the ninth day after cutting the polyp, I thought I saw three small protuberances emerging from the edges of the anterior end of the second part, the one which had neither head nor arms. The moment I saw them I thought of the horns which serve the polyps as arms and legs. These protuberances were precisely where the arms would have been had this second part been a complete polyp. I did not want to conclude so quickly, however, that these were indeed arms that were beginning to grow. I continued to see these protuberances throughout the day, and I became extremely excited and impatient for the moment when I would know clearly what they were. Finally, on the next day they were large enough to dispel all doubt; these were truly arms growing at the anterior end of the second part. On the following day two new arms began to emerge, and a few days later three more came out. This second part then had eight arms which, in a short time, were as long as the arms of the first part, that is, those arms the polyp had before it was cut up.

From that time on I found no difference between the polyp that developed from the second part and a polyp that had never been cut up. The first part had given me that impression since the day following the operation. When I examined the two parts through a magnifying glass with all the attentiveness of which I was capable, both of them appeared to be demonstrably complete polyps performing all the functions of these organisms known to me: they elongated, contracted, and took steps.

My experiment thus yielded results far exceeding my expectations. According to the original reasoning behind this experiment, however, I would have had to conclude positively that the polyps were plants, and moreover, plants which could grow from cuttings. Nevertheless I was very far from hazarding such a decision. The more I observed whole polyps, and even the two parts in which the reproduction just described took place, the more their activity called to mind the image of an animal. Their movement seemed spontaneous, a characteristic always regarded as foreign to plants, but one with which we are familiar through endless examples among animals. Everything I had done to extricate myself from doubt had served only to plunge me deeper into perplexity. Consequently, I resolved to redouble my watchfulness and to try to discover in the polyps some property which would characterize them more definitively.

All who have attempted to compare plants and animals have recognized the great similarities existing between these two classes of organisms and the difficulty of specifying precisely the particular characteristics that distinguish one from the other. The reflections of others on this subject, added to my own, served only to confirm my uncertainty about the polyps and impelled me to search for new properties in these organisms likely to resolve my doubt. Something else of which I had learned recently also greatly affected my thinking: the discovery made on the aphids.

Mr. Réaumur had long suspected that these little animals could reproduce without having mated since their birth. In May, 1740 Mr. Bonnet undertook to ascertain whether this supposition was true. The results achieved through Mr. Bonnet's care and astuteness were presented in the thirteenth Memoir of Mr. Réaumur's *History of Insects*, Volume Six. Mr. Bonnet found and proved by experiments performed with all the necessary precautions, that a young aphid would reproduce although it was kept in absolute solitude from the first instant of its birth. In July of that same year, Mr. Lyonet attempted the same experiment on the aphids as did Mr. Bonnet, and met with the same result. I learned the outcome of Mr. Bonnet's experiment from one of his letters, and Mr. Lyonet, in his study, showed me his aphids that were kept isolated from one another and that were reproducing. I had the pleasure of seeing that these two gentlemen, unknown to each other, had performed the same experiment although on aphids of different species; almost simultaneously they had discovered one of the most remarkable facts of natural history, one which ran directly counter to the rule generally accepted until that time on the reproduction of animals. I was curious to repeat the experiment of these gentlemen. At the very time that my first sectioned polyps were showing the exceptional reproduction which I have described, I also had some aphids kept isolated from one another from birth, that were reproducing.

Such a phenomenon as these aphids exhibited could not help but inspire me with considerable distrust for generalizations. In particular, it increased my distrust of rules which would rank the two properties I had found in the polyps under two different classes of organisms, one characteristic belonging to animals and the other to plants. I felt keenly that nature was too vast and too little known for us to decide without temerity that such and such a property was not to be found in such and such a class of organisms. I reserved judgment and was satisfied to push forward with research on the polyps without venturing to conclude as yet whether they were animals or plants.

At that time I did not know how the polyps multiplied. I thought that perhaps such knowledge could provide the distinctive characteristic I was seeking and enable me to judge whether they were animals or plants. I had a large jar heavily stocked with green polyps, and I often spent hours at a time examining them one by one. Finally, I found a polyp that was beginning to produce a little one. At the end of several days, my observations on this polyp led me to recognize that the manner by which these organisms multiplied was very closely akin to the way plants multiply when they give off *shoots*. This property constituted yet another characteristic of plants that the polyps had just shown me. Nevertheless, I remained very much inclined to consider the polyps animals, or rather plant-animals, since they seemed to belong to a species midway between these two classes of organisms.

Thus, the new characteristic that I had just discovered left me no less doubtful than I had been for so long beforehand. In spite of my impatience to know precisely in which class the polyps belonged, I nevertheless experienced some pleasure in this doubt. It had already led me to two of their very remarkable properties, and it was stimulating me to search for others by piquing my curiosity more and more.

At that time, I had just sent a second shipment of polyps to Mr. Réaumur because those in the first had died. The great naturalist sent me news of their safe arrival in Paris, word I had been awaiting impatiently. In this message he included a decision which alone sufficed to put an end to all my doubts. After examining the polyps, he did not hestitate to classify them as animals, and he gave them the name which they now bear because they resemble the polyps of the sea [that is, squids and octopuses].

It was in March, 1741 that Mr. Réaumur resolved my uncertainty. In the following month, I discovered a property in the polyps which would have convinced me had I still been in doubt.

At that time, I found a new species of polyps (Plate 1, Fig.2), different from the kind I had been observing for a long time. A few days after finding these new polyps, I saw them eat. I saw them swallow worms as long and even longer than themselves, and I saw that they digested them and that they were nourished by them. These observations certainly provided convincing evidence that they were animals.

Now that I have given the history of the discovery of the major unusual trait that I found in the polyps, I am going to relate, in the order which seems most natural to me, the observations I have made on these animals during more than three years of studying them.

I have said beforehand that the first polyps I saw were fastened on aquatic plants which I had taken from a ditch and put into a large jar full of water. Among them were some duckweed, one stalk of a kind of horsetail, and a water lily. At the beginning of my studies I searched for polyps mainly on these plants, but I subsequently learned that they fasten themselves indiscriminately to any object in the water. I have found them on all kinds of water plants, on the bottom of ditches, and suspended at the surface of the water; I have seen them on branches of trees, on boards, decayed leaves, fragments of straw, and on stones. Finally, I have even seen some on the bodies of various animals, for example, on snail shells and on the cases of caddis worms.

To detect polyps upon such objects as I have just mentioned while the polyps are still in the ditch requires an experienced eye and knowledge of how to choose the proper location and conditions. The most convenient method is to take the various objects out of the ditch, and to put them into glass jars full of water where the polyps attached to them may then be easily seen.

In order to come to know an animal, it is very useful to observe it in its natural conditions, that is in the midst of everything that surrounds it in the places where it is found. For this reason it is desirable that the vessel in which the animal is kept be arranged very much like its original habitat. This expedient can hasten considerably the research one is projecting; it can even lead to findings which otherwise would not be made.

In delving into the natural history of the polyps, not only did I have recourse to this expedient, but in addition I took care to return frequently during the summer to the edge of the ditch in which I had found them. I would go at the hour when the sun shone through to the bottom of the water, and I would choose places where the water was clear and the bank had a gentle slope. I could see the polyps distinctly at the bottom of the water, on all the objects that were in the water, and at its surface. Through this method I acquired ideas which I never would have gained without taking such precautions.

The most common position in which the polyps are found, be it in their natural habitat or in the vessels where one is keeping them, is that shown in Figures 1 and 2 of Plate 1. The posterior extremity b of the body of the polyp $a\,b$ is fastened against a plant $e\,f$ (Fig. 1) or against the piece of wood $g\,h$ (Fig. 2); the body $a\,b$ and the arms $a\,c$ are extended out in the water.

The most usual shape of the body when in this position is not exactly the same for all of the three species of polyps with which I am familiar. The body of the green polyps (Fig. 1), my principal subject until now, tapers a bit from its anterior to its posterior end. This tapering is almost imperceptible. The same is true of the second species of polyps shown in Figure 2. But the body of the third species (Fig. 3) differs in this respect from that of the preceding two: it tapers imperceptibly from the anterior extremity *a* only as far as the halfway point *d*, and occasionally to a point two thirds (Fig.4, *d*) of the way down the length of the body. In that area the body becomes much thinner, and from there to the posterior extremity it no longer tapers. This part (Figs. 3 and 4, *d b*), more slender than the rest of the body, has the look of a tail and can serve as a distinguishing characteristic of this species.

As I have already said, the arms, or legs, are at the anterior extremity of the polyp. Each arm can move in all directions, staying in line with the body or forming all sorts of angles to it. Furthermore, each arm moves independently of the others. When all the arms remain at the same angle to the body, and when they maintain the same alignment as they stretch away from their points of origin, they present an extremely symmetrical appearance. This appearance changes as the angle formed by the arms and the body changes. In the green polyps the arms and body are at right angles, or thereabout (Fig. 1, *i* and *k*). Then the arms, issuing at equal intervals from the anterior end of the polyp and continuing in a straight course, all appear to emanate from a common center like the radii of a circle (Fig. 1, *i*). But when the arms happen to draw closer together frontwards, forming an obtuse angle with the body (Fig. 1, *a*), they take on the shape of a funnel which may be more or less flared.

Seldom do any but the green polyps show so symmetrical an alignment of their arms. Because their arms are short, they can hold them straight more easily. In polyps of the two other species known to me (Figs. 2 and 3) the arms are much longer. When extended they continue in a straight line from their points of origin only for a certain distance. The remaining portions of the arms go in other directions, sometimes uniform among all of them but often quite diverse.

I have already mentioned that not all individuals within the species of green polyps have an identical number of arms. The same is true of the other species which I subsequently observed. In the three species of polyps with which I am familiar the smallest number of arms is commonly six and the greatest is twelve or thirteen. I have seen some polyps of the second species, however, which had eighteen arms (Plate 10, Fig. 3).

A question arises here which first occurred to me as I was making these observations. Is it natural for the number of arms to vary among the individuals within a species? Or rather, have those with fewer arms lost some accidentally?

In the foregoing I have already responded to this question in passing. I remarked that even when I was only slightly acquainted with the polyps, I considered it natural that they did not all have an equal number of these horns, or rather arms and legs (as they were renamed once they were better understood). After observing the polyps for some time and especially after seeing their birth and growth, I realized that their arms do not develop as do the analogous parts of so many other animals we know. The arms and legs of most animals appear all at the same time from the moment that they begin to develop. The arms of the polyps, to the contrary, grow successively with new ones developing even long after birth. This fact brings to light in a very obvious way one reason for the unequal number of arms found among the different individuals within the same species of polyps.

I was not content to stop there. It was possible that although the arms of polyps do not all develop at the same time, there might be a natural number of arms for the species which each polyp would eventually attain. My observations, however, taught me precisely the opposite. I have not found any pattern in the increase of the number of arms of the polyps, and I have never seen them reach a uniform number.

Now I return to the body of the polyps shown extended in Figures 1, 2, 3 and 4 of Plate 1. Like the bodies of many well known animals, it is capable of extending and contracting to varying degrees. The body of a polyp can contract to the point where it is only about two millimeters long (Plate 1, Figs. 5 and 6). For example, the polyp illustrated in Figure 3 of Plate 1 can contract to the size of the one shown in Figure 6. The size of a contracted polyp varies depending on the species and on the size of the particular polyp. Its body, whether in the act of contracting or extending, can stop at any conceivable stage from the fullest extension to the tightest contraction. Consequently a polyp can vary the length of its body to an extreme degree. If it were necessary for the polyp to extend as fully as possible, or even to a particular length, in order to carry out certain movements, as is the case with other animals, it would be possible to ascribe an exact length to it by indicating the length it has in this situation. Except while it is extremely contracted, however, a polyp can execute the complete repertoire of its movements and maneuvers regardless of its length. Thus, the length of the polyps' bodies can be described only in approximate terms, all the more so because of the individual variations in growth shown by each polyp of a species, as is natural both in animals and in plants.

The majority of green polyps that I have seen are between eleven and fourteen millimeters in length when extended. Polyps of the second and third species are usually between eighteen and twenty-seven millimeters long, but I have seen some of these two species with bodies that measured an inch and a half (Plate 10, Fig. 3).

There is no need to belabor the point that the body of the polyp becomes proportionally thinner as it elongates and thicker as it contracts. Figures 1, 2, and 3 of Plate 1 depict the thickness most commonly observed in full-grown and normally extended individuals of the three species I have discussed.

Since many of the animals capable of lengthening and shortening their bodies have body rings, it was natural at first to look for them in the polyps. This I did, but neither the magnifying glass nor the microscope enabled me to see any regardless of whether the polyps were contracted or extended when I observed them. Their manner of extending and contracting appeared to me related more to that of snails and slugs than to that seen in worms and other creatures which have visible body rings. Figure 1 of Plate 5 is an exact delineation of a polyp seen enlarged through a microscope. The observer who drew it would not have failed to show rings, if he had found any, nor would they have escaped his attention.

I will not stop here at all to explain the means by which the body of a polyp contracts and extends. Were I to do so, I would risk offering nothing but conjectures.

The extent to which polyps contract depends on how roughly one touches them or shakes the water surrounding them. Every polyp taken from the water comes out as a small contracted mass fastened against the object to which it has attached itself (Plate 1, Fig. 7). It looks so different from an extended polyp that it is hard to recognize at first. Once the eye is accustomed to seeing contracted polyps out of the water, however, it becomes easy to single them out from all other objects. Such an experienced eye is very useful when looking for polyps because it eliminates the need for putting the various objects that harbor them into water in order to make the polyps assume a more recognizable shape.

Heat and cold affect polyps as they do the majority of terrestrial and aquatic creatures. Heat makes them active, and cold makes them sluggish. Only when the temperature drops very close to freezing, however, do the polyps become completely inactive. Then the polyps are more or less contracted and remain so. As the water surrounding them warms up a few degrees, they elongate. They resume all their customary movements in proportion to the rise in temperature. It is not necessary that the water become very much warmer in order for the polyps to elongate a great deal; it need only approach a moderate temperature, that is, 48 degrees on the Fahrenheit thermometer. In summer they elongate still further and more frequently; but the effect of several degrees of heat more or less is not clear enough to be noted precisely.

The arms of the polyp extend and contract in the same manner as does the body. Because the arms, like the body, can stop at any stage between full extension and complete contraction, any specification of their length is, up to a certain point, ambiguous. Nonetheless, it is easy to determine that polyps of a particular species have longer arms than do those of another. Such is true, at least, of the three species with which I am acquainted.

The green polyps have the shortest arms (Plate 1, Fig. 1). Rarely have I seen any that would exceed half the length of the body, that is their arms were no more than seven millimeters long. Arms an inch long are very common in polyps of the second species (Fig. 2). I have seen a number of them extend up to two and even three inches.

I have already stated that polyps of the third species are easily recognized by their tails (Fig. 3 *a b*), but the length of their arms provides an additional conspicuous feature useful in distinguishing them from the other species. For this reason, I shall often call them *polyps with long arms*.

I have named the polyps which I am discussing the first, second, and third species according to the order in which I found them. I discovered the first species, the green ones (Fig. 1), in June 1740; the second (Fig. 2) in April 1741; and the third or polyps with long arms (Fig. 3) in July of the same year.

When I was withdrawing some of the polyps with long arms from the water, I mistook them for polyps of the second species. Hastily I thrust great numbers of them into a jar of water measuring seven inches in height and five in diameter; I then placed the jar on a window sill. Not until the following day did I have an opportunity to examine these polyps carefully. I was greatly surprised to find the jar filled with extremely long threads which appeared to me as fine as spider's silk. I first set out to investigate these threads without suspecting what they really were. Soon I realized with astonishment and true pleasure that those long fine threads came from the anterior end of the polyps; in a word, they were arms. Because the jar contained many polyps, it was richly decorated with their arms, some arranged nearly in a straight line, and others winding about and making all sorts of twists and turns. Like the arms of other species, they taper from their bases to their tips. At their base they are no thicker than are the arms of polyps of the second species. Toward their tips the extended arms of the third species are more slender than those of the second species, just as they are longer. I have seen some polyps of the third species with arms up to eight and a half inches long. The effect produced by these arms when they are well extended can be appreciated by viewing Figure 3 of Plate 1 which depicts them as in life.

It is easy to understand that in order to see these arms elongate to the extent I have just described, the polyps must be in large vessels. I put some into a jar in which their arms could have extended to a length of ten inches in a straight line, but I saw them reach only eight and a half inches (Plate 1, Fig. 3). I do not wish to conclude, however, that they are incapable of elongating further.

I have compared their fineness to that of spider's silk. If the arms are not quite so fine at their extremities as the threads of a spider, they nevertheless come very close. In the illustrations included in this work it has not been possible to make these arms appear as fine as they actually are, nor to represent very clearly in the polyps drawn lifesize how these arms taper from their bases to their tips. This fineness and tapering may be seen distinctly in the polyp shown enlarged as seen through the microscope (Plate 5, Fig. 1). Later on I shall have occasion to return to a discussion of these long arms, which have furnished me a number of highly entertaining and curious spectacles and which have excited the wonderment of all who have seen them.

The first time I saw the green polyps contract, their arms disappeared completely from view. At first I thought that they had withdrawn into the body of the polyp as do the feelers of snails. After studying the anterior end of the polyp carefully with the aid of a magnifying glass, however, I recognized its arms and saw that they had not retracted into the body at all, but were only strongly contracted. Later I discerned them with the naked eye. Albeit they were not as contracted as when I first looked at them, but by that time my eye was accustomed to seeing them. When strongly contracted the arms of polyps of the second and third species measure between two and five millimeters in length (Plate 1, Figs. 5 and 6).

Ordinarily the movement or touch which causes the body of a polyp to contract produces the same effect on the arms. The arms also contract on their own to carry out all the maneuvers, which I shall discuss later, related to their mode of locomotion and their manner of seizing, holding, and carrying prey to their mouths.

There is not always a correlation between the contraction and elongation of the bodies and the arms of the polyps. When one makes the body contract by touching it a bit roughly, both body and arms contract considerably; but in other situations the body contracts considerably without the arms doing so, and vice versa. Neither do the arms and body necessarily move in concert upon elongating once again.

The arms of an individual polyp can extend and contract, fully or in part, independently of each other. Often some will be very long while others are very short (Plate 1, Fig. 4).

When stimulated by some external cause green polyps contract rapidly, whereas those of the two other species do so more slowly. When acting spontaneously, however, all polyps contract fairly slowly. Elongation is always slow, although the rate varies. These considerations apply both to the polyp's body and to its arms.

I come now to another characteristic of both the body and the arms of the polyps about which I have already said a word in passing: their flexibility. In fact, all parts of the body and the arms of the polyp are capable of bending in every direction and to every possible degree. Figure 1 of Plate 2 portrays some of these particulars; polyps are found in such postures. The body (Plate 2, Fig. 2) and arms (Plate 1, Fig. 4) as well are capable of twisting into loops. One notes a place along the arms of polyps of the second and third species where they ordinarily bend. It is located about five millimeters from the point of origin of the arms. Each arm of a particular polyp can perform all manner of flexions independent of the movements of the other arms.

Polyps of the third species usually let their arms trail downward, often turning and twisting in a variety of ways (Plate 1, Fig. 3). When these polyps are at the bottom of large glass jars, however, often I have seen them point some of their arms toward the top, at times perpendicularly, and at other times I have seen the arms as they are drawn in Figure 3 of Plate 2. This figure very accurately depicts a polyp which I saw in one of my large jars.

The prodigious variety of shapes and postures that these animals can assume is evident from what I have just said about the degrees of extension, contraction, and inflection of which they are capable. Recall that such is the case with the body of the polyp, or its arms in general, or each of them individually. In addition, one can often see the same polyp assume a number of different shapes in the course of a single day. It would be useless to describe them here. The illustrations included in this work provide sufficient examples.

I should also add that polyps can inflate their bodies, sometimes in one place, sometimes in another, and often in several places at once. I have seen some of their bodies embellished with such a number of encircling bulges that they resembled quailpipes [instruments used to attract quails]. If polyps were observed only in this condition, these bulges could be mistaken for annular body rings.

The polyps move forward by means of their ability to extend, contract and bend in all directions. Let us consider the polyp *a b* in Plate 3, Figure 1, secured by its posterior end *b*, while its body *a b* and its arms spread out in the water. To advance it bends and brings its anterior end *a* near the surface on which it is traveling. Next it fastens against that surface, sometimes with only its anterior end, at other times with only a few arms, and at yet still other times with both the arms and the anterior end *a* (Fig. 2). Once the anterior end has been attached securely, the polyp detaches its posterior end *b*, moves it close to the anterior end *a* and takes hold with *b* (Fig. 3). Then it again detaches the anterior end *a* and extends it anew (Fig. 4). There you have in general the description of a step as taken by a polyp.

Such an account may suggest that this manner of locomotion closely resembles that of various aquatic and terrestrial animals, such as the inchworms and some species of fairly common aquatic worms. The inchworms just mentioned, which are depicted in the work of Mr. Réaumur (Volume 1, Memoir 2, Plate I, Figs. 13, 15, and 16), merely bend their bodies in order to bring the posterior end closer to the anterior, whereas the aquatic worms I mentioned both bend and contract their bodies. Polyps likewise both bend and contract without there always being the same relationship between the degree of inflection and the degree of contraction. At times they contract much more than they bend, and vice versa. The variations in these actions are so great that they contribute greatly to the consequent variations in the polyps' steps. Furthermore, polyps do not always advance their posterior end an equal distance toward the anterior.

When polyps walk they perform these movements very slowly. Often they will pause in the midst of a step and move and twist their body (Plate 2, Fig. 2) and their arms in all sorts of ways. At times polyps take steps of quite an extraordinary nature. Of these I shall describe only two, those that impressed me because they appeared far different from their usual manner of walking.

Consider now the polyp *a b* shown in Plate 3, Figure 5, fastened by its posterior *b*, with its body and arms extended in the water. In order to perform one of the extraordinary steps in question, it will first move its anterior end *a* toward the surface on which it is traveling and will attach itself there by *a* (Fig. 6). Next, it will detach its posterior end *b* and raise it straight up, making its body perpendicular to the surface (Fig. 7). After that it will bend its body down on the other side and will fasten its posterior end at *b* (Fig. 8). Then it will detach its anterior end *a* and straighten its body once again, this time with the anterior end uppermost (Fig. 9, *a*). Such a step, or a somersault rather, is executed in truth so slowly that while the polyp is still carrying out one, an agile tumbler could perform a great many.

Now for the other extraordinary step I mentioned. Consider the polyp *a b* shown in Plate 3, Figure 10. It is fastened by its posterior end *b* against the wall of a jar, its body and most of its arms stretching forward. One of its arms *a c* holds fast to the glass at *c*. When the polyp is in this position, it detaches the posterior end *b* and slightly contracts the body so that the posterior end is brought a little nearer to point *c*. This end is then fastened to the glass at *d*. Thereupon the polyp repeats the same maneuver and attaches its posterior end at *e*. I have sometimes seen the posterior end detach itself and then, after contracting the body slightly, attach itself again three and four times in succession.

All that I have said about the locomotion of polyps applies equally to the three species known to me. Anyone who has developed a taste for the study of small creatures cannot help but enjoy watching the polyps as they perform all the movements which I have just described.

Polyps travel at the bottom of the water; they ascend along its edges or onto aquatic plants; and often they come up to the surface of the water where they hang suspended by their posterior end (Plate 1, Fig. 4, *b*). I have even seen some hanging there by a single arm (Plate 2, Fig. 4, *c*), but this position is exceptional whereas the other is quite common. Polyps walk on the water's surface from underneath it as they do over the solid objects I have previously described.

By watching the polyps move in a jar, one can get an idea of the movements they make in open water. When taken from the water, polyps pass from the plants or other objects to which they were attached, onto the bottom or the sides of the vessel into which they are placed. They ascend along the sides until they reach the surface of the water and continue to move along under this surface until they stop there, or they walk across the surface proceeding to the other side of the jar over which they then travel.

Because polyps take their steps very slowly and often with a considerable lapse of time between steps, they spend a long time traveling a short distance. Judging from the great number of polyps I have kept in jars, seven or eight inches is a good day's journey in summer for the polyps. When it is cooler they move even more slowly and consequently advance less. Of the three species known to me, the green polyps travel the most rapidly. Although they move slowly, one might say they were fast in comparison with the polyps of the two other species.

One can determine by what means a polyp remains at the surface of the water by using a magnifying glass to examine attentively the posterior extremity of one of the animals suspended there (Plate 1, Fig. 4, *b*). This extremity is out of the water, it is dry and it is at the base of a little hollow (Plate 1, Fig. 4, *b*, and Plate 3, Fig. 11, *b* and *c*). The extremity itself constitutes the bottom of that hollow, and the water forms its edges. To be persuaded that this dry condition is absolutely indispensable for the polyp to be maintained at the surface, one need only moisten its dry extremity with a drop of water and immediately the animal will fall to the bottom. A polyp holds on at the surface of the water by precisely the same method as would be used to keep a pin there. One places a pin on the water with care to avoid getting it wet, and then it will remain on the surface. The side of the pin that does not touch the water remains dry, lying in a hollow formed by the water itself. The heavier the pin, the larger and deeper the hollow will be.

Thus, to pass from the side of a jar to the surface of the water, a polyp need only project out of the water ever so slightly the part by which it will hold on at the surface and allow it time to dry. That is just what it does and what an observer can easily see if he seeks the opportunity. For example, consider the polyp *e f* (Plate 3, Fig. 11) which is fastened against the side of a jar near the surface of the water. To get to the surface, it raises its anterior end *e*, projects it above the surface and lets it dry for a moment. Next it detaches its posterior end *f* from the glass and brings it above the surface. In a moment its posterior end is dry and able to support the polyp which then draws its anterior end under the water. Thus the polyp comes to be suspended at the surface of the water. Frequently it elongates its body and arms right away.

What I have just described may suffice to explain how polyps travel near the surface of the water and also how they leave it to pass onto the sides of a vessel or onto the objects which are in the water.

In the course of experiments I have carried out on the polyps, it often was necessary for me to have the ones on which I was working suspended at the surface of the water. Since I could not always be sure that they would go there on their own, I sought a means of suspending them there at my will. At first I tried several expedients in vain, and I did not succeed until I had a clear idea of how polyps are able to maintain themselves at the surface of the water. Once I saw that their posterior end was kept dry outside the water, I undertook to dry the posterior extremity of the specific polyps that I wished to have at the surface of the water. This method proved successful.

For this purpose, I begin by placing them in a shallow vessel and I wait until they are well extended. Then I take them out of the water with the aid of an artist's brush, a very useful tool that is needed continually in handling the polyps. I place the end of the brush under the polyp, slowly push it towards the surface of the water, and lift it out upon this brush to which it remains attached. Usually, when touched and pushed by the brush, the polyp contracts at least partially, completing the contraction at the moment that it leaves the water. It is desirable, however, that it not be extremely contracted. One can successfully prevent too great a contraction, first, by waiting until the polyp is well extended before attempting to remove it from the water and, secondly, by performing slowly and delicately whatever needs to be done to place the polyp onto the brush.

I must further advise that as the polyp comes out of the water, it must lie with its anterior extremity on the tip of the brush. The reason for this procedure will become evident. The polyp on the brush must remain out of the water for a minute or so. Then I take the brush in one hand and a pointed quill in the other. Little by little, I immerse the tip of the brush in the water and with it the anterior extremity of the polyp which is attached to it; I continue to lower the brush until only a part of the polyp's posterior extremity, about a millimeter long, remains outside the water. At that moment, using the tip of the quill held in my other hand, I remove from the brush that part of the polyp which is already in the water and which often has already detached itself as it floats. By this means and by blowing against the polyp, I detach the posterior extremity, which is still touching the brush and which is outside the water. As soon as it detaches, I withdraw the brush and leave the polyp undisturbed. The end which is outside the water usually remains there, the body elongates in the water, and the polyp turns out to be suspended at the surface. This procedure does not always succeed; a number of mishaps can cause the attempt to fail. Nothing then remains but to begin again.

If a polyp is already suspended at the surface of the water and one wishes to change the water, as is often necessary, it is easy to do so. In this case one suspends the polyp at the surface of the new water in which one wishes to place it and which must be in another vessel. It is necessary to place the brush parallel to the polyp and then move it closer until it touches the polyp. The polyp fastens itself against the brush, is drawn out of the water, and its posterior end remains dry so that it can be placed at once in the fresh water using the same precautions pointed out a moment ago.

However lengthy the description of the method I use to suspend polyps at the surface of the water, I believed it necessary to present it for the sake of those who may wish to repeat my experiments and to perform new ones. In this way, I spare them the time and the trouble that they would spend trying to find strategems and I leave them in no worse a position to devise better ones.

I have never seen polyps swim, and it appears they are not able to. I have detached polyps that were in all kinds of circumstances and in various degrees of extension and contraction from the objects on which they were fastened; I have removed others from the surface of the water where they were suspended; I have placed others in the middle of the water. Not one made the slightest motion to swim. All of them fell to the bottom of the water more or less speedily, depending on the degree to which they were contracted or extended.

I have given the names of arms and legs to the filaments projecting from the anterior extremity of the polyps. From my description of the polyps' manner of locomotion, it may be seen that in fact these parts do function as legs. When I describe the process whereby the polyps seize and hold their prey, however, the reader will readily conclude that the name arms suits them especially well.

Let me indicate here yet another function of these arms, that of clinging tightly at times to the objects on which the polyps are fastened. Usually the polyps are attached only by their posterior end, sufficing to prevent their being dragged away by the motion of the water, even when it is considerable. A number of times I have observed polyps which not only were fastened to the bottom of my jars by their posterior ends, but which in addition had two or three arms stretched toward and attached to different sides of the glass. The polyp secured in this manner could not be tossed about by the movement of the water. Possibly at times such an arrangement may be useful to these animals when they are at the bottom of ditches. There is one situation, however, in which it must be even more important for them to use their arms as anchors and cables to avoid being carried away by the movement of the water. That is the case when they are suspended at the surface exposed to every motion of the water. Let us suppose that a polyp, especially one with long arms, is suspended at the surface of the water, and that it is expedient for it not to be carried away by the movement of this water. It could cast anchor by means of its arms, fastening their ends against the bottom of the ditch or against the plants and other objects in the water. Were it to fasten three or four arms on different sides, it would be precisely in the position of a ship moored by its anchors. I am not presenting this at all as an indisputably obvious function of the arms; it is merely a conjecture based upon one of my observations.

Two polyps with long arms were suspended at the surface of the water in one of my jars (Plate 3, Fig. 11, *a b* and *d c*). One day I found them in a position closely resembling the one I have just proposed. One polyp *d c* had two of its arms *d i* and *d k* fastened by their extremities against the bottom of the jar on two different sides *i* and *k*. The other polyp *a b* had one arm *a g* also secured to the bottom of the jar and another arm *a h* attached at its tip *h* to the side of the glass rather close to the bottom. I blew on the water to agitate it a bit, but the waves I raised did not carry the polyps completely away. They were kept within a certain area by their anchors, that is by their arms, which were stretched taut and fastened firmly to the glass. I had to shake the jar vigorously to make them detach.

From all I have said, it must by now be evident that polyps are able to attach themselves firmly to the objects on which they rest. I have performed some experiments which may provide an idea of how strongly they adhere. The precautions required to maintain polyps will provide those who wish to raise them with many opportunities to witness the facts I wish to describe.

In order to keep these animals healthy, their water must be changed quite frequently, especially after they have eaten, as will become clearer from what I say later on [*Memoir* II, p. 83]. If one wishes to change the water around a polyp fastened to the walls of a powder jar which does not need cleaning, one need merely toss the water out. No matter how roughly the water is poured out, it will not dislodge the polyp which will remain attached to the glass. The strength of the polyp's grip becomes even more evident when fresh water is poured into the vessel. Ordinarily neither the pressure of large amounts of water falling from a height of several inches, nor the turbulence caused by the impact on the water already in the powder jar, suffices to detach the polyp. Its anterior end is simply swept about in every direction by the turbulent water, but the posterior end does not disengage from the glass. One readily imagines the strength of the adhesion required to withstand such stress.

This adhesion is voluntary. As described earlier, a polyp moves by alternately detaching and attaching its arms and both extremities of its body.

It is therefore able to control how firmly it adheres to a surface. The question is whether a polyp exercises this control by exerting an effort or by some other means. The Polyp is too small an animal to permit experiments to be made that would answer this question conclusively.

With his usual sagacity, Mr. Réaumur has conducted such experiments on the adhesion of the limpet. He determined some of the essential factors involved in the adhesion of this mollusc and then applied what he learned to the limbs of starfish and sea anemones. He found that such animals attach themselves to objects primarily by means of a viscous secretion. The adhesion of the limpet results from the effect of this glue combined with the meshing of the parts of the animal's skin with the irregularities on the surface of the objects to which it adheres. This naturalist discovered that the limpet, to detach itself, moistens the base of its body with a liquid that it exudes. This liquid dilutes the glue which was holding the limpet fast and takes the place of the force that would be necessary to overcome the adhesion. "The base of the animal," says Mr. Réaumur of the limpet, "appears covered with a countless number of granules, calling to mind the appearance of shagreen leather. It is quite certain that parts of these granules consist of little vesicles filled with liquid, for when one makes an open wound in the base, no matter how slight, the liquid escapes from them. The viscous substance or glue in question is contained in another part of the same granules. Or, alternatively, some other vessels spread the glue through the base of the limpet. To attach itself, the animal expels glue from the vessels that contain it and presses its base, thus moistened, against some rock which the sea has left uncovered as it receded. To leave the rock, it has no need to use force. As we have shown, that force would be equivalent to thirty pounds of weight. The limpet needs merely to squeeze the vesicles containing the water; the water escapes and dilutes the glue and the animal is at liberty to go in search of suitable food." (*Mém. de l'Acad. Roy. de Sci.,* for the year 1711, p. 113.)

It is quite apparent that Mr. Réaumur's words on the causes of the adhesion of limpets also apply to that of the polyps. However smooth the surfaces, such as glass, upon which polyps fasten, there remain enough irregularities into which they can press extremely small bits of their skin. This meshing, combined with the effects of the viscous substance, may serve to explain the adhesion of the polyps, though I would not wish to allege, nonetheless, that no other causes may be involved.

It seems certain that the polyps contain some viscous substance. When placed dry on the hand and handled, they seem to be composed exclusively of such matter; it feels as though one were handling a bit of slime.

It is certain that the body of a polyp resembles shagreen leather. Indeed it is abundantly covered with tiny granules. I would not wish to conclude, however, that the function of these granules or their only function consists of supplying the viscous matter and the liquid the polyps need to attach or detach themselves, or to seize what their arms encounter, or to release what they have grasped.

I have already referred to polyps of the first species many times by their color, calling them the green polyps. Those of the two other species, when taken from the ditches where they grow, usually are a reddish brown. The polyps with long arms at times are nearly a flesh color.

It is impossible to be precise in specifying the color of the last two species because it varies greatly. Even when the polyps of one species are taken from a ditch at the same time, they are far from sharing the identical shade of color. Various hues of brown and red are to be found, and occasionally one will come upon polyps of a completely different color. I have carried out decisive experiments upon this subject of colors. To avoid being repetitious, however, I feel obliged to delay this account until I come to the section discussing the food of the polyps, with which the subject of colors is closely related [*Memoir* II, p. 77].

The shade of color also changes according to the degree of extension or contraction of the polyps. The more extended the polyp, the lighter the shade; the shade becomes much deeper when the polyp is tightly contracted. This is the case with the three species with which I am familiar.

The following observation also applies to these three species of polyps: they can lose their color and become white and then regain their color again. They lose and regain their color gradually, accounting in part for the differences in hue found in the various individuals of the same species and in the same polyp at different times. The variation in hue to which I am referring here is independent of the changes resulting from the varying degrees to which the polyp contracts or elongates.

When examined under the magnifying glass and the microscope, the surface of a polyp's body appears shagreened as if covered with tiny granules (Plate 5, Fig. 1). The surface looks this way both when the polyp elongates and when it contracts. The extent to which the surface appears shagreened can vary as the body elongates and as other conditions change.

As can be discerned even with the naked eye, the color seen in the polyps is not located on their surface at all. Especially when aided by a magnifying glass or a microscope lens, the observer will be able to recognize that the polyps are covered with something transparent and that it is only what lies underneath that gives the polyps their color. This transparent envelope is very clearly depicted in Figure 1 of Plate 5. I must caution here, however, that it is best not to think of this transparent envelope as a distinct skin separate from what is underneath.

The anterior end of different polyps does not seem always to take the same form. As I have already stated, and as the figures which I have cited illustrate, their arms project from the edges of this anterior portion. It is the area between the arms that I wish to discuss, for this is the part that does not always seem to be shaped the same way. Very often it is elongated, appearing as a small conical protuberance (Plate 1, Fig. 3, *a* and Plate 4, Fig. 4). At times, the cone it forms seems truncated (Plate 5, Fig. 1, *a*). At other times, no protrusion can be discerned, the space between the base of each arm appearing completely flat (Plate 2, Fig. 2, *a* and Plate 4, Fig. 5). Finally, in other circumstances this area is concave; the anterior end of the polyp is open and is shaped a little like the mouth of a bell (Plate 1, Fig. 1, *i*, Fig. 2, *e* and Plate 4, Fig. 6). This last circumstance, however, is not the only one in which the anterior end is open. Using a magnifying glass, one can sometimes observe a small hole in the anterior end when it lies flat and also when it is in the shape of a truncated cone. The hole is shown enlarged in Plate 5, Figure 1, *a*.

At this time I shall avoid going into detail regarding the function of the anterior end which, as I have shown, is able to open and close. As my present aim is merely to describe the polyp in general terms, I shall note only that the opening at this end serves it as a mouth. Thus their arms grow out from their lips (Plate 4, Figs. 4, 5, and 6), and it is the form taken by the lips while extending and contracting that determines the different shapes I noticed in the anterior end, or to say it otherwise, in the head of the polyps. Now, after all I have just said, I can call it a head.

The occasion arises here to compare freshwater polyps with polyps of the sea and to justify the name given the former by pointing out the similarity of their external structure to that of these marine animals. The arms or legs of the marine polyp [that is, octopus or squid] are situated at its head and around its mouth. We can now conclude that such is also the position of the arms or legs of the freshwater polyp. Both use these members in locomotion, and I shall soon show that the freshwater polyp, like the marine polyps, makes use of them also to seize prey and carry it to its mouth.

The mouth of our polyp opens into its stomach, a sac which extends all the way from the mouth to the posterior extremity. This cavity can be seen with the naked eye but is especially clear under a magnifying glass. To see it clearly, one must expose the polyp to broad daylight or to the light of a candle. The color of the polyp is not sufficiently opaque to prevent the observer from discerning quite clearly that the polyp is tunneled through from one end to the other. This inner cavity is portrayed in the figure of a polyp as seen enlarged under the microscope (Plate 5, Fig. 1, *a b*).

To convince myself that polyps have a cavity running from end to end, I was not satisfied to observe just the outside of the polyp but sought to see the opening by looking at the inside of it. To accomplish this goal, I cut a polyp transversely into three parts, each of which contracted to an extremely short length. I placed all three parts at the bottom of a flat glass vessel full of water and then saw very clearly that they had a cavity from one end to the other. Looking down through the upper end of each section, I could see at its lower end the glass on which each part rested (Plate 4, Figs. 1, 2, and 3), clear proof of their hollowness. Since each of the three parts was hollow from end to end, it follows quite obviously that the polyp which they had formed had likewise been hollow from end to end. The mouth at the anterior extremity of the first part (Plate 4, Fig. 1, *a*) was at this time wide open. The posterior extremity of the polyp was at the end of the third part (Fig. 3, *b*). The fact that it was hollowed through from one end to the other showed clearly that this end of the polyp also can open. There are fewer opportunities to see this end open than to see the mouth open. A number of facts I report later on should serve as additional proof that the polyps are hollow from end to end.

It is easy to understand how greatly and in what manner the shape of this canal can vary, considering that it extends the length of the polyp's body, and that it is affected by the changing shape of the polyp as it elongates, contracts, twists, bends, and swells at different spots along its body. The canal at times appears cylindrical and at other times wider or narrower in some spots than in others (Plate 5, Fig. 1). In the polyp with long arms the body tapers considerably at a point about two-thirds down its length and the canal narrows proportionally (Plate 5, Fig. 1).

I have called this opening which extends from one end of the polyp to the other a stomach, for it is there indeed that the food is carried and digested. Often it is full of water which can enter easily because the mouth is almost always open.

Occasionally I have found polyps which had a bubble of air in their stomachs. The first time I saw one of these bubbles, I doubted that it really was one. There was an easy way to find out. Any polyp that is not holding fast somewhere falls by its own weight to the bottom of the water. The one with the stomach containing what I thought was a bubble of air was attached against the side of a powder jar. I had only to detach it, and then, if it had no bubble, it would fall to the bottom. If it did have an air bubble, it would be lifted to the surface of the water by the effect of this bubble, as happens with all the polyps which have a bubble on the outside of their posterior extremity. I therefore detached my polyp; it rose at once to the surface of the water.

The skin which encloses the stomach of the polyp and which forms this sac that is open at both ends is the skin of the polyp itself. The entire animal consists of nothing more than a single skin, shaped in the form of a tube, or gut, open at both its ends. When some other animals such as caterpillars or sundry species of worms are cut open, they are found to contain various ducts. On opening a polyp, however, the observer finds absolutely no more than one canal extending the length of the entire body. In other words, as I have already said, this entire animal seems to form but a single vessel, the external surface of which is the surface of the animal itself.

I have said that one finds but a single vessel in the polyp. By this I meant only that I have not been able to discover any other. It may be that there are some others within the skin of the polyp which may be so small that they have escaped my scrutiny and have not been discerned.

Clearly, this cavity extending from the mouth to the posterior extremity is the food canal, the stomach of the polyp, where food is initially prepared to become nutriment. I will have occasion shortly to demonstrate that the canal serves this purpose [p. 75]. In the skin forming this stomach (Plate 5, Fig. 1, a b), there must be parts that subsequently receive the nutritive substances. In addition, this skin must contain not only all the organs necessary for the nutrition and growth of the polyps, but also those needed to carry out their movements. The nature and arrangement of these parts, however, are bound to be extremely difficult to uncover in an animal as small and soft as a polyp.

By analogy one could conjecture that there are such and such imperceptible parts in the skin of the polyps, but I find it hard to believe that these suppositions founded on simple analogy would be very satisfying. The polyps have a variety of characteristics which are directly contrary to analogies drawn from so many other animals. May they not also differ from those animals in regard to the nature of the unseen parts of which they are composed, and in the animal economy which results from the structure and functions of these parts? This possibility seems more than likely to me.

In order to try to discover what the polyp's skin is composed of, I laid it out under the microscope in all kinds of ways. One of my first concerns was to inspect the edges, that is, the borders of the wounds of a polyp cut transversely. With this aim in mind, I placed a section of a polyp on a piece of glass in such a way that I could look directly at the cross section of the skin where the polyp was cut (Plate 4, Fig. 2, *a*). It appeared to me that what I called the transparent surface of the polyp, the envelope enclosing the colored portion, was tightly united to it and was just like it in the arrangement of its parts.

These parts that I wish to discuss and that will be mentioned frequently later on are tiny granules which, when seen through a magnifying glass or a microscope, seem to line the edges of the skin of a piece of polyp that is being examined. This skin is full of these little granules throughout its entire thickness. These are the same granules which, when observed on the external surface of a polyp, make it appear shagreened.

I could not help being curious to know whether or not similar granules existed on the interior surface of the skin of the polyps, on the walls of their stomach. Accordingly, I opened a number of polyps. Here is how I go about it.

I place a polyp on my hand and make it contract as much as possible. Next I insert into its mouth one blade of an extremely sharp scissors, pushing it out through the posterior end. I then close the scissors, thereby cutting one side of the skin of the polyp along its whole length and opening the inner canal from one end to the other. After that, by pressing the severed skin down on all sides, I expose the interior surface of the skin of the polyp, that is, the walls of its stomach. The polyp itself is now nothing more than a simple skin spread open with its external surface laid against my hand and its inner surface facing upward (Plate 4, Fig. 7). At this juncture, with the aid of a strong magnifying glass, it is easy for me to examine the interior surface of the polyp at leisure. If I wish to look at it under the microscope, I need only transfer the skin onto a glass slide.

By using a magnifying glass of very short focus and a microscope with strong lenses, I was able to discover in this inner surface of the skin of the polyps (Plate 4, Fig. 7, *a*) a large quantity of those same granules that I had already seen in their external surface and in the cross-section of a piece of skin. The inner surface appeared even more shagreened than the exterior and much less smooth. Instead of being transparent, as is the exterior surface, it had a tinge of the color of the polyp.

I have compared these two surfaces using a large number of polyps opened expressly for this purpose or for other experiments. As will be seen later [p. 160], it was rather important for me to have made the comparisons. As I continue to narrate my findings on the structure of the skin of the polyps, I shall have occasion to indicate in greater detail how these surfaces are alike and how they differ.

It seemed to me that the skin was made up entirely of these tiny granules that are seen distinctly on both surfaces and on the edges of polyps that have been cut (Plate 4, Figs. 2, *a* and 7, *b*). To examine a piece of skin in more detail, I put it in a drop of water that was on a glass slide and then laid it out under the magnifying glass and the microscope (Plate 4, Fig. 8, *a*). I have already mentioned that large quantities of granules can be seen around the edges of a piece of a polyp's skin. I drew out great numbers of granules by pressing the skin with the point of a pen, by rubbing it against the glass, and by trying to tear it (Fig. 8, *b, c, d*). The granules dispersed in all directions in the water and then came together in clusters (Fig. 8, *e, f*). When I gently tilted the slide that held the granules, little currents developed which swept them along (Fig. 8, *b, c, d*). These currents drew up along the edges of the piece of skin from which the granules had come (Fig. 8, *a*), detaching more of them. These observations led me to conclude that these granules must not be strongly attached together. The current of water frequently separates them out and drags them apart.

With all possible care I sought to determine whether the granules lay enclosed within vesicles, but I was not able to find any. When one places torn pieces of polyp in a drop of water, the granules are separated out by that water. Thus, in a whole polyp there must be something that holds the granules together and thereby prevents water from separating them out.

The remarks I am about to make concerning the viscous material of the polyps perhaps will indicate one factor accounting for the coalescence of these granules. Let me refer again to the pieces of a polyp's skin placed in a drop of water on a slide. As I tore the pieces of skin and stirred them about in the water, I examined them through a lens of very short focus. I saw a number of granules which appeared joined together in a viscous substance.

It is certain that polyps are viscous; one will be convinced readily of this fact with one's own eyes, even without observing the polyps attentively. When I lift up the point of the quill (Plate 4, Fig. 9, *a*) sticking into a piece of the skin of a polyp (Fig. 9, *b*), it sometimes draws along a strand of viscous matter which often is lined with granules. Through a strong microscope lens, I conveniently saw this viscous material spin out from a piece of polyp as I pressed down on its skin with the point of a quill which was split as are those used for writing. The cleft in the quill opened with the pressure, and then I could see a strand of viscous substance, varying in thickness, pass through the opening (Plate 4, Fig. 9, c). These strands are usually lined with granules, but by dint of stirring and shaking a small piece of skin in a drop of water, I made most of the granules detach from the viscous substance, so that the strand remained at the end of my quill almost free of them. During this procedure I examined the specimen continually through a strong lens.

There is thus no reason to doubt that these granules of the polyps are mixed with a viscous substance which tends to hold them together. Shortly, in discussing the structure of the arms of the polyps, I shall present evidence of the cohesiveness of this substance. It seemed to me that there was always more of it in the exterior surface of the polyps than in the interior one, thus making the outside appear smoother than the inside. Indeed the polyps often appear enveloped in this material which serves them in some way as skin.

When one lays out an intact polyp under a microscope without injuring it, usually one does not fail to see granules that have detached along some areas of its surface. This is to be noted in the healthiest of polyps; when these granules detach in large numbers, however, it is a symptom that the polyp is dangerously ill: The surface of the affected polyp becomes increasingly irregular. It no longer has the finish it had previously. Granules escape from all sides. The polyp contracts and swells up as do its arms (Plate 4, Fig. 10, a). It takes on a whitish look. Finally it loses its form altogether, and in the place where the polyp was, nothing more than a pile of granules remains (Fig. 10, b). One could say that the viscous substance which held the granules together lost its cohesiveness and from that time the granules became detached. In order to see the effects of this malady clearly, one should observe it in a polyp of the second or third species, preferably one of the largest.

Here I need to note one other very important observation concerning the granules with which the skin of a polyp is embellished. When a small portion of skin is torn in a drop of water on a glass slide, and a large number of granules are forced out, some collect together in little piles in various locations within the drop; others remain singly. Seen through a microscope, these little piles of granules appear to be the color of the polyp from which they were drawn. This coloration is especially obvious in the granules drawn from the green polyps because their green color is brighter and more striking than the reddish brown of the other polyps. It is plainly noticeable, nevertheless, that the granules of the two last species are colored. When the granules of any of the species are not gathered together, however, they are a transparent white; for example, after one separates those that appeared colored when heaped together, they appear transparent.

There is nothing extraordinary in this fact. Small parts of a colored object, when separated, do not appear as vividly colored as when they are joined together. Further on [p. 81] I shall disclose facts which perhaps will help to explain why this phenomenon is more noticeable in the granules of the polyps than in bits of many other colored materials.

Because the granules of a polyp when joined together are the same color as the polyp, no doubt the color of the polyp resides in those granules and is dependent upon them. I said above (p. 26), however, that the polyps can lose their color and become pale. Does this mean, then, that they lose their granules, and does the loss of color result from the loss of these granules? The question is certainly interesting, and one senses that the answer would illuminate the nature of the polyps' granules. To throw light on this problem, one had only to observe carefully whether those polyps which lost their color totally or in part still had granules, or at least if they retained but a few of them, to observe whether a proportional loss of color and granules had taken place. This I did.

I observed a number of pale polyps; to make my observations more conclusive, I also inspected some colored ones at the same time. I found an abundance of granules in the pale polyps, many more it seemed to me than there should have been if the numbers of granules had decreased in proportion to the diminution of the polyp's color. From this observation one may conclude that the polyps do not lose their granules in proportion as they lose their color, and consequently that it is the granules themselves which lose this color and cease to be colored.

Here is one more fact which will complete the evidence. The little piles of granules drawn from a polyp that has lost its color are not colored at all, whereas all the piles of granules from a colored polyp are themselves always colored in proportion to the degree of color of the polyp and the quantity of granules gathered together.

I have dwelt at great length on the subject of the granules because my study of them has furnished me with the only ideas I have about the constitution of the polyps. These ideas, although indistinct and imperfect, seemed interesting to me, and perhaps they will appear so to others once they have considered what I have yet to say on this matter. For example, I have succeeded in discovering what causes the coloring in the granules of the polyps, but I feel obliged to postpone all the detail until I have explained some of the facts relating to the nourishment of the polyp and their consequences [*Memoir* II]. When the reader comes to the later observations regarding the granules, it will be very important for him to recall what I have said thus far.

The structure of the arms of a polyp bears a strong resemblance to that of its body. Examined under a magnifying glass or a microscope, the arms show a shagreened external surface both when extended and contracted, as does the body of the polyp (Plate 5, Fig. 1). A strongly contracting arm will appear extremely shagreened, more so even than the body. It appears less so proportionally as it elongates, and when well extended it does not appear shagreened over its entire length. At such times I note a considerable variation in the same arm. This phenomenon that I wish to discuss is seen in polyps of the second and third species, but especially in those of the third species because their arms elongate greatly. I will devote my attention here to describing the arms of this latter species.

When it is contracted, the surface of an arm appears extremely shagreened all over, richly bedecked with tiny granules. The amount of granulation on the surface changes continually in proportion to the degree to which the arm extends; this change is more noticeable near the end of the arm than at its base (Plate 5, Fig. 1).

Imagine an arm as it extends. Little by little one sees these surface granules, which touched or nearly touched each other when the arm was strongly contracted, become spaced wider apart (Fig. 2). When the arm reaches a certain degree of elongation, its surface appears merely sprinkled with beads (Fig. 3). Drawing further away from each other as the arm continues to elongate, the beads finally turn out to be strung in a single row and separated by a transparent thread (Fig. 4 *i, i, i*). These beads are formed by the clustering of a number of granules. At first glance they appear like so many beads threaded on a string so that they do not touch. When one watches more attentively, however, one perceives that the thread does not pass through the center of these beads at all (Fig. 4).

The various stages through which an arm passes as it elongates, as I have just described, can be seen all at once in different areas of the same arm (Plate 5, Fig. 1), because the arm does not elongate equally throughout. Ordinarily the tip of the arm extends first and these beads I have mentioned appear quite separated from each other, whereas in the area near the base of the arms the beads hardly seem to have been moved apart at all. My statement can be verified by looking over the entire length of the arms of the polyp shown enlarged in Figure 1 of Plate 5, but especially by paying attention to the portions of the arms shown extremely enlarged under the microscope. Figure 2 represents an elongated arm as it appears close to its base with the granules barely separated. A greater separateness is depicted in Figure 3, which shows the middle portion of an extended arm. Finally, Figure 4 shows the granules set out in a row as they appear mainly near the tip of the polyp's arm. This tip often ends in a bead (Plate 5, Fig. 1).

The hairlike structures drawn in Plate 5, Figures 3 and 4, *e, e, e* become noticeable in the elongated arm of a polyp when it is examined under the strong lens of a microscope. They appear to be transparent.

To observe the arms of polyps, especially those of the third species, I make use of glass slides as follows: with the aid of a magnifying lens, one may easily observe an arm, whether elongated or contracted, through the glass wall of a powder jar containing the polyps. To arrange for such an opportunity, it suffices to have a number of polyps in a glass container. Some will always be found with their arms close enough to the sides of the powder jar to be within range of a magnifying glass that has a focal length of eleven to fourteen millimeters. To examine the arms under the microscope, however, they must be taken out of the water. Such a procedure presents no difficulty when one intends only to examine an arm while it is contracted, because as the polyp is taken out of the water its arms contract. Then one merely cuts an arm from the polyp and places it on a glass slide that can be adjusted to a microscope.

What demands the greatest of care, however, is the fixing of an elongated arm on that slide. To achieve that end, I select a polyp which has been fastened high on the side of a powder jar very close to the surface of the water. At the moment that one of its arms is well extended, I take in one hand a quill at the end of which is a brush, and in the other hand I hold between two fingers one end of a glass slide two or three inches long and about twelve millimeters wide. Next I bring the point of the brush into contact with the tip of the elongated arm that I wish to observe. When the tip attaches itself to the brush, I draw it very gently out of the water and the rest of the arm follows it. In this way I have a portion of the arm out of the water. If it has not sufficiently elongated, I can make it extend as I draw it out. The polyp fastened against the jar remains in place while its arm is being pulled, forcing the arm to either stretch or break. It stretches, and enough so that if handled delicately, it can be made to elongate greatly.

All that remains to be done in order to put the arm on the slide is to maneuver the slide under this arm until they touch. Then, by making a small motion with the hand that holds the slide, I break the arm off at both edges of the slide. A portion of the arm remains on the tip of the brush, another on the polyp; the middle section is fastened against the slide. It can be examined under all kinds of microscopes. It does not change at all when it dries; at least it remains for several days just as it was when first taken from the water.

To detect the color in a polyp's arm, one must examine it attentively; otherwise it will appear pale. By observing it under a magnifying glass and a microscope, especially when it is contracted, one discovers from the base up to a certain distance, a tinge of the same color as was evident in the polyp of which it was a part.

Because the arms of the polyps are the same color as is the body, we are led naturally to another similarity between the arms and body. We have seen that the body of the polyp takes its color from the granules that cover it. It is therefore quite natural to presume that the arms also have such granules. To be convinced of this fact, it suffices to tear off some arms on a glass slide and to examine them under a microscope. It is also these granules that make the surface of the arms appear shagreened and, as the arms elongate, come together to form the little beads that I have mentioned. These beads contain a number of these granules (Plate 5, Figs. 2, 3, and 4).

The material in which the granules are contained, and which I have termed viscous when speaking of the polyp's body, can be seen very distinctly in an elongated arm. When it is sufficiently elongated so that the beads are spaced some distance apart in a row, the thread that separates them (Plate 5, Fig. 4, *i,i,i*) is composed solely of this material. Its transparency is easily determined by examining it under a magnifying glass or a microscope. To gain some idea of the cohesive quality of this material one need only consider the capacity of a polyp's arms to resist strain without breaking. Later on, there will be a number of opportunities to assess the strength of these arms [*Memoir* II, pp. 55 and 64].

My comments about the arms may help a bit to clarify our still extremely confused ideas regarding the structure of these animals. It is quite apparent that the polyps' strength lies in this viscous substance in which must be contained all the components that serve to carry out their movements of contracting, elongating, flexing, and so forth.

We have yet to learn whether the arms of a polyp are hollow inside as is the body. I venture to affirm that they are, and I intend to offer clear proof. But the evidence on which this proof depends will come forward more naturally at another juncture. At that time I will also establish that the cavity inside the arms is connected with the stomach of the polyp.

I have been unable to cut open an arm in order to see its inner surface as I have seen that of the body. It is to be presumed that their surfaces are very similar.

All that I have written thus far should serve to facilitate the understanding of what is to follow in these *Memoirs* concerning other aspects of the natural history of the polyps, for example, the manner in which they are nourished. Likewise the observations on the nutrition of the polyps will furnish me the opportunity to delve a little more deeply into the subject of their structure.

At the start of this *Memoir*, I mentioned the propensity of polyps for light and showed that they always go to the best-lit side of the jar. I was not satisfied, however, simply to have them assemble on the best-lit side. I encased a large jar well-stocked with green polyps within a cardboard sleeve, as within a muff. On one side of the sleeve I had cut an opening in the shape of a chevron. When the jar was enclosed in the sleeve, this opening corresponded to the middle of that side of the jar. When I arranged the jar so that the aperture in the cardboard was turned toward daylight, the polyps always gathered on the side of the jar corresponding to the opening, their assemblage taking on the form of a chevron juxtaposed to the one that had been cut in the cardboard. I turned the jar inside its sleeve very frequently, and during the period of several days I invariably saw the polyps forming a chevron at the opening. To further vary the experiment, I arranged the cardboard so that the chevron lay sometimes pointing upward, sometimes downward; the cluster of polyps assumed the corresponding inverted or upright form.

The supposition may arise that, rather than seeking light, the polyps seek the most air and the warmest air. However, even when I arranged the opening in the sleeve so that it was exposed to the coldest air rather than to the warmest, the polyps gathered to no less a degree near this opening. If it were the greatest amount of air that they are seeking, they would all collect at the surface of the water.

It therefore seems certain that it is indeed light that attracts these animals to the best-lit area of the jar. This is not a new discovery; various terrestrial and aquatic animals that move towards light have been known for a long time. Several kinds of flies and nocturnal butterflies afford constant examples of this phenomenon on summer evenings. Few people are unaware of the use made of candlelight and torches as lures to attract and catch various aquatic animals. This method is used, for example, to catch marine polyps on the coasts of the Adriatic Sea. They come to the light that fishermen shine in the dark on the surface of the water.

In addition, there are various species of small aquatic creatures which seem to have a marked propensity for light. Among them I know of one kind which, according to its structure, even should be classified among the polyps.

I became curious to trace the exact route taken by the green polyps as they proceeded towards the opening in the cardboard sleeve that I have been discussing. It often happened that more than a hundred polyps would be clinging to the bottom of the jar as I encased it in the sleeve. The majority began by ascending along the wall of the jar on the far side away from the opening. From there they advanced to the surface of the water. Crossing the surface, they then passed to the side of the jar closest to the light and continued until they were situated against the opening. If the polyps could swim, the shortest route to the opening would be to swim directly to it; but, as they cannot, they are obliged to proceed along the glass walls and the surface of the water.

It is to be noted that in making this journey which brings them to the brightest location, the polyps pass through dark areas, for example those parts of the jar not reached by the light from the opening. This observation made me think of investigating whether they continue, even after nightfall, on the route towards the best-lit area. For three evenings in succession, between eight and nine o'clock, I marked the location of twenty green polyps. They were ascending the side of the jar farthest from the light in order to continue over the surface of the water to the best-lit area. I placed a little piece of paper against the jar next to the posterior end of each of these polyps. To make certain that they could not avail themselves of even the most feeble rays of light, I sealed up the opening with the greatest care for the night. Thus I could be sure that they had not seen any light at all even if I could not come to observe them before daybreak. I performed this experiment in the winter. Each morning on the three occasions when I had marked their positions, I came to view my travelers and not once did I find more than four polyps that had remained stationary. Some had advanced more than an inch on their journey. Those which had been near the surface of the water the evening before had passed onto it, and the others had ascended along the wall of the jar to which they had clung the previous evening when I had marked their place.

I have kept the jar which I just discussed in a sleeve that had no opening, and I never saw the polyps in it assemble in any one location.

Since the green polyps are more active than those of the other two species, they make better subjects for the experiments I have just described. Nothing is easier to ascertain, however, than that the others also seek the light. To do so, it is sufficient to keep a number of them for some length of time in the same jar without altering its position.

EXPLANATION OF
THE FIGURES OF THE FIRST MEMOIR
PLATE I

FIGURE 1 depicts three green polyps, lifesize, fastened by their posterior ends to a duckweed plant *e f*. Polyp *a b* is seen from the side, its arms making an obtuse angle with its body; *k d* also is seen from the side, but each of its arms makes a right angle with its body. Polyp *i g* is shown with its anterior end forward and its mouth *i* open.

Figure 2 illustrates two polyps of the second species attached by their posterior ends to a piece of wood *g h*. Polyp *a b* is shown from the side; polyp *d e* has its body curved towards the front, showing its open mouth at *e*.

Figure 3 portrays a polyp of the third species, lifesize, with its arms hanging down and greatly elongated. I frequently have seen arms at least as long as those depicted in this figure. The head *a* of the polyp terminates in a conical protuberance, a shape often assumed by the lips of the polyps. At *d* the polyp narrows, so that *d b*, its posterior section, seems to be a sort of tail, much more slender than the rest of the body.

Figure 4 shows a polyp of the third species suspended by its posterior end at the surface of the water. At *b* is the small hollow formed at the surface of the water, at the bottom of which lies the tip of the posterior end of the polyp. The polyp narrows at *d*. The arms of this polyp vary greatly in length. It was drawn in this way in order to show that these arms do not always elongate equally; in fact it is very rare to find them the same length. Two of the arms in this figure are twisted about near their base.

Figure 5 represents a polyp of the second species with both its body and arms greatly contracted.

Figure 6 shows a polyp with long arms, the arms and body of which are also greatly contracted. In this position the narrow posterior part of the polyp, marked *d b* in Figure 3, is discernible; it always remains more slender than the rest whether the polyp is extended or contracted.

In Figure 7 a water lily leaf is seen out of the water and turned upside down. Three rather large polyps are revealed clinging to the lower surface of the leaf, just as they look when taken out of the water.

PLATE 2

FIGURE 1 represents a polyp of the third species, its body and arms bending in various directions. These inflections may serve as examples of some positions the polyp can assume; it would not be possible to illustrate all of them. Other examples may be seen in many of the other figures included in this work.

Figure 2 portrays a polyp of the second species attached by its posterior end *b* to a piece of wood, its body twisted in the shape of a posthorn [a single looped instrument of the bugle family]. Polyps at times may be seen taking this shape or a considerable number of others similar to it. Looking at the head *a*, one sees that the lips are not extended to form a cone but lie joined together and flattened so that neither a peak nor a hollow is visible.

Figure 3 depicts a polyp of the third species fastened by its posterior end *b* to the center of the bottom of a large powder jar. Its body *a b* stands erect on this base. Two of its arms extend mostly toward the top of the container, but their anterior portions fall back down toward the bottom. The other arms frame the body on all sides, their tips falling to the bottom of the jar. This figure represents a polyp that I saw in which all these different postures were combined at the same moment.

In Figure 4 a polyp of the third species hangs suspended from the surface of the water by one of its arms at *c*. I have seen polyps in precisely the position portrayed here.

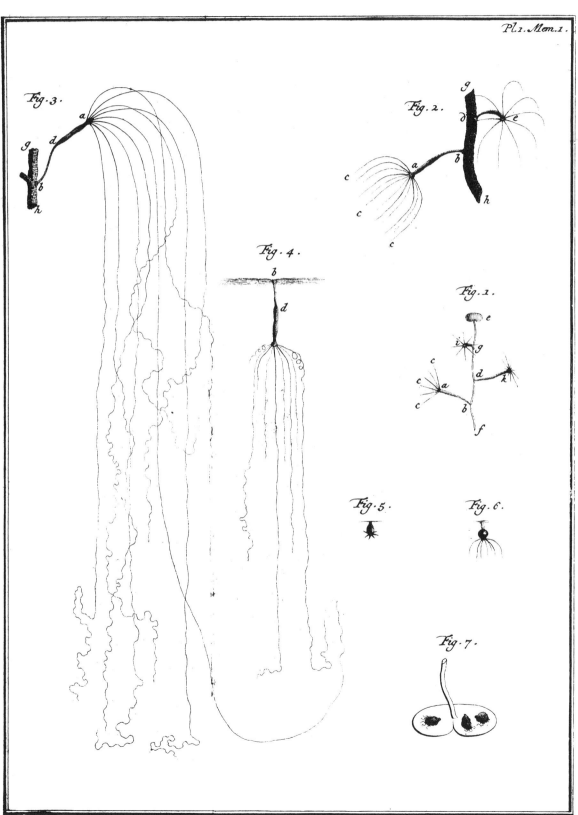

Fig.3.

Fig.2.

Fig.4.

Fig.1.

Fig.5.

Fig.6.

Fig.7.

PLATE 3

FIGURES 1, 2, 3, and 4 show the same polyp, but in the various positions it assumes in taking an ordinary step. In Figure 1 the posterior end *b* is fastened to an object of some kind, the body *a b* standing partially erect. In Figure 2 the polyp *a b* of Figure 1 has bent its body and attached itself by its anterior end *a*. After the anterior end is fastened securely, the posterior end *b* detaches and draws close to the anterior end *a*, as seen in Figure 3. In this position the body of the polyp is curved much more than in the preceding figure, its two ends nearly touching each other. It often happens, however, that the posterior end *b* does not move as close to the anterior end *a* as is shown in Figure 3. Figure 4 depicts the polyp of Figure 3 once it has detached its anterior end *a*, raised its body partway and extended forward. Thus the polyp completes its step. Afterwards the polyp either remains still, its form more or less extended, or else it begins another step. It executes the movements described here extremely slowly, often allowing considerable intervals to elapse between movements.

Figures 5, 6, 7, 8, and 9 depict a single polyp in the various postures it assumes in performing an extraordinary kind of step. In Figure 5 the polyp stands nearly perpendicular to the object on which it is fastened by its posterior end *b*. In Figure 6 the same polyp has bent its body around and attached itself by its anterior end *a*. Now it detaches its posterior end *b* and raises its body up as shown in Figure 7 so that its posterior end *b* is uppermost and the polyp is standing erect by its anterior end on the surface to which it clings. Next this posterior end *b* curves downward, and the polyp bends in the opposite direction from that shown in Figure 6. Figure 8 depicts the polyp in this position, its posterior end *b* once more attached. Finally, the anterior end *a* is detached and the body is restored to an upright posture. Figure 9 shows this position, the posterior end *b* fastened to the object, the anterior *a* raised upward.

Figure 10 illustrates another extraordinary step taken by the polyp. The polyp is shown attached at *b* by its posterior end to the wall of a jar. The arm *a c* is likewise attached to the wall at *c*. In order to move towards *c*, the polyp in this situation detaches its posterior end from point *b*, contracts its body a little, and then fastens its posterior end anew at point *d*. A moment later, it again detaches this end and fastens it at *e*. Thus it has covered the entire distance between *b* and *e*.

Figure 11 shows a jar filled with water up to *l m*. It contains the three polyps *e f, d c,* and *a b*. Let us suppose that at first the polyp *e f* was simply attached at *f* by its posterior end, its arms and body extending in the water. In order to proceed to the surface of the water, it raised its anterior end *e* until it came ever so slightly out of the water at *e*. Once this end is dried, the polyp detaches its posterior end *f*, and bends its body *e f* until the posterior end is brought to the surface of the water and raised above it. This done, it can then pull its anterior end back down under the water and extend its arms and body. It will hang suspended at the surface of the water by its posterior end, as do polyps *d c* and *a b*. Tiny hollows are formed at the surface by the posterior ends of these polyps at *c* and *b*. Polyp *d c* has its arms *d i* and *d k* strongly attached by their tips to the walls of the jar at *i* and *k*, and polyp *a b* has its arms *a g* and *a h* also strongly attached by their tips to the walls of the jar at *g* and *h*. I have observed two polyps in a single glass situated precisely as are the two drawn in this figure. They were, so to speak, moored by their arms as is a ship by its anchors. These four arms, serving as cables and anchors, were stretched in straight lines as shown here with *d i* and *d k*, and *a g* and *a h*.

44

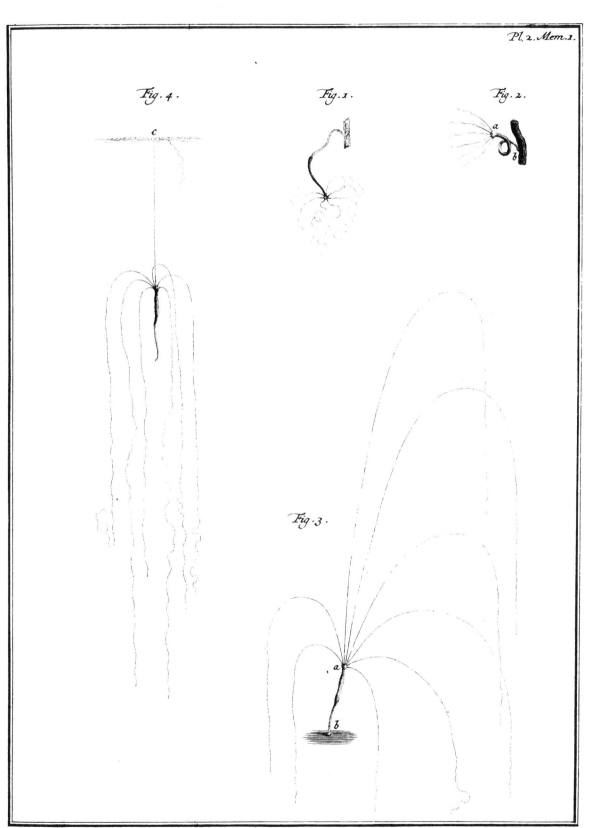

Pl. 2. Mem. 1.

Fig. 4.

Fig. 1.

Fig. 2.

c

a

b

Fig. 3.

a

b

P. Lyonet delin.

J. v. Schley sculp. 1743.

PLATE 4

FIGURES 1, 2, and 3 represent three parts of the same polyp sectioned transversely. Light can be seen through each of these sections, just as it can be seen through a straight tube open at both ends. Since each of these three parts is pierced through with a cavity from one end to the other, it proves that a similar cavity extends from one end to the other of the whole polyp. In Figure 1 the first section is seen resting on its anterior end *a*. The mouth which forms the extremity is wide open, and the arms are strongly contracted. Figure 2 presents the second section of the polyp, the letter *a* indicating the thickness of the skin as it appears in such a sliced part. Figure 3 shows the third or last section of the polyp resting on its posterior end *b*.

In Figures 4, 5, and 6, three heads of polyps are depicted as seen enlarged through a microscope. The figures show more distinctly the various contours that the lips of a polyp can assume; these contours affect the shape of the anterior extremity of the polyp. In these figures the arms are drawn showing them cut very close to their bases so as not to obstruct the view of the mouth. The remaining portions of the arms are shown open at their anterior tips. Although I have never seen them like this, nevertheless they must appear so the moment after the cuts have been made. I base this statement on proof to be given later [p. 160] that the arms of the polyps are shaped like a hollow tube. Figure 4 portrays the mouth of a polyp with lips extended forward in the shape of a cone. This shape also can be seen lifesize in Plate 1, Figure 3, *a*. Figure 5 shows a head with flattened lips, as was represented lifesize in Plate 2, Figure 2, *a*. Figure 6 represents a polyp with the anterior end open and flaring outward; this contour can also be seen lifesize in Plate 1, Figure 1, *i* and Figure 2, *e*, and in Plate 3, Figure 1.

Figure 7 shows a polyp which has been opened lengthwise from one end to the other as it appears enlarged through a microscope. The walls of the stomach, that is, the interior surface of the skin, are at *a* whereas *b* represents a section of that skin. Figure 8 represents a piece of skin *a*, as it appears enlarged under the microscope, when it has been placed on a slide in a drop of water. Granules *b, d,* and *c* have come out from the skin and are shown drawn along in the currents formed by the water when the glass slide is tilted. Small piles of granules are seen in *e* and *f*. Figure 9 depicts a quill *a*, its tip pressed against a piece of skin *b*. The tip is divided like the point of a quill used for writing. The slit is open, and a thread of viscous material *c* is seen stretching across it.

Figure 10 portrays a polyp that is seriously ill and that is in the process of losing its shape completely as it dies. Nothing then remains of it but a small heap of granules *b*, shown here enlarged.

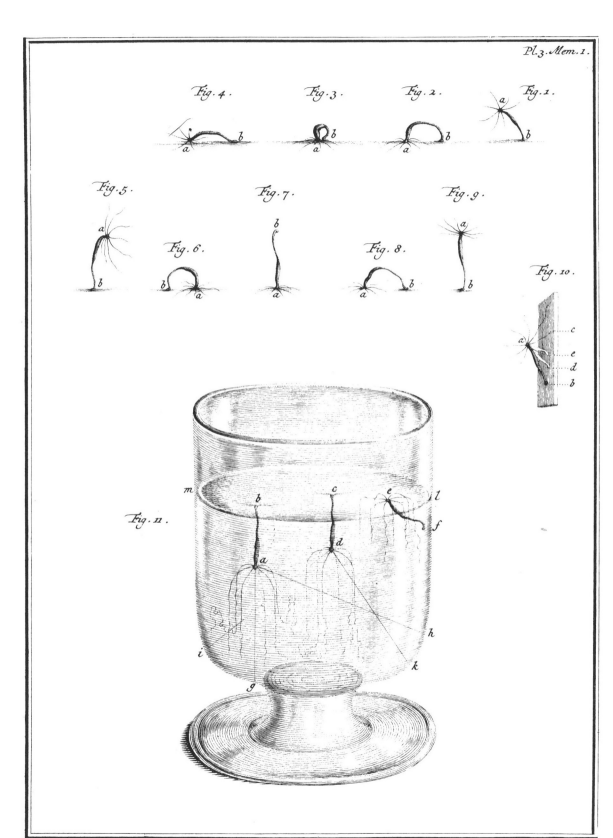

Pl. 3. Mem. 1.

Fig. 4.

Fig. 3.

Fig. 2.

Fig. 1.

Fig. 5.

Fig. 7.

Fig. 9.

Fig. 6.

Fig. 8.

Fig. 10.

Fig. 11.

P. Lyonet delin.

J. v. Schley sculp. 1743.

PLATE 5

FIGURE 1 presents a polyp of the third species as it appears enlarged through a microscope; the posterior end *b* is attached to a piece of wood. The cavity *a b* that extends from the mouth at the anterior end *a* to the posterior end *b* can be seen with great clarity. Polyps rarely have necks as pronounced as the one depicted here, although those who observe many polyps and who observe them frequently will have occasion to see them. Considering that the arms of this species may attain seven or eight inches in length, those drawn in this figure are not nearly as long as they should be had they been shown enlarged in proportion to the body. Thus it is necessary to assume that the arms here are not depicted sufficiently elongated. I have often seen polyps with arms arranged almost like those in this figure. The engraving conveys perfectly the shagreened appearance of the surface of the body and the arms.

Figure 2 represents a piece of a polyp's arm near its point of origin as it appears when partly elongated and as seen enlarged through a powerful lens of the microscope. Although I have never seen the end *o* open in the manner shown in this figure, undoubtedly one could see it open if one could only observe the cross-section through a microscope the instant it was cut away. The empty space within the arm of the polyp near its point of origin is depicted with great precision as seen through a microscope. It can also be perceived through a magnifying glass. This cavity is wider at end *o* than at the other extremity, because *o* marks the beginning of the arm, from which point the cavity becomes progressively narrower. The granules that give the skin of the polyp its shagreened appearance are seen here lying quite close together.

They are somewhat farther apart in Figure 3 which shows a more elongated section of an arm than seen in Figure 2. A kind of transparent hair *e, e, e* can be found on the extended arms of polyps examined through a strong lens of the microscope. Figure 4 represents the end of an extremely elongated arm. The granules extend in a single row with intervals between them at *i, i, i* and appear to be attached to a transparent thread. At points *e, e, e* are hairs that one can discern through the microscope.

Pl. 4. Mem. 1.

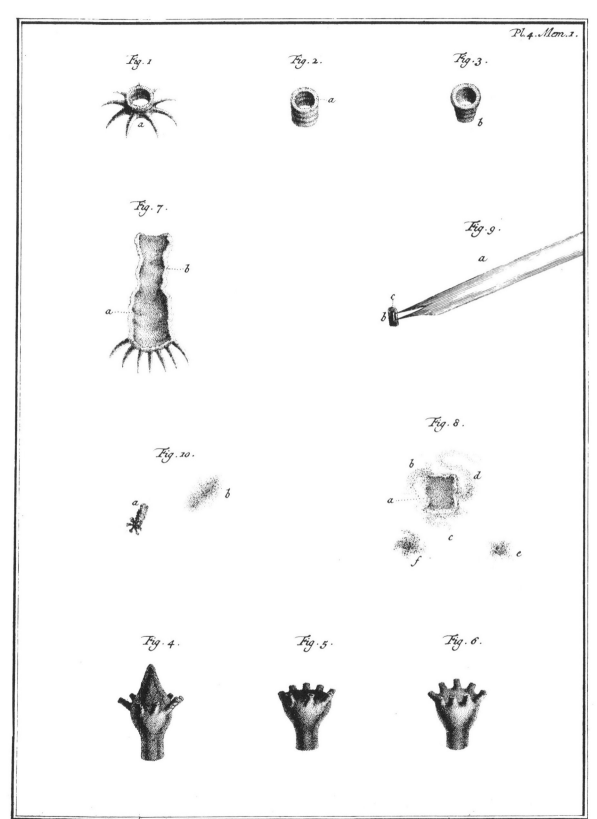

Fig. 1.

Fig. 2.

Fig. 3.

Fig. 7.

Fig. 9.

Fig. 10.

Fig. 8.

Fig. 4.

Fig. 5.

Fig. 6.

P. Lyonet delin.

J. v. Schley sculp. 1743.

C. Pronk del. ad Viv. 1744.   J. v. Schley sculp.

MÉMOIRES
POUR L'HISTOIRE
DES POLYPES.

❊❊❊❊❊❊❊❊❊❊❊❊❊❊❊❊❊❊❊❊❊❊❊❊❊❊❊❊❊❊❊❊❊❊❊

SECOND MÉMOIRE.

De la Nourriture des Polypes, de la Maniére dont ils saisissent & avalent leur Proie, de la Cause de la Couleur des Polypes, & de ce qu'on a pu découvrir de plus sur leur Structure. Du Tems, & des Moïens les plus propres pour trouver des Polypes.

O N a vu dans le Mémoire précédent, que ce sont les Polypes verds que j'ai trouvés les premiers. Je les ai observés pendant plus de six mois; & quelques soins que je me sois donnés pour découvrir comment ils

L   se

Pl. 5. Mem. 1.

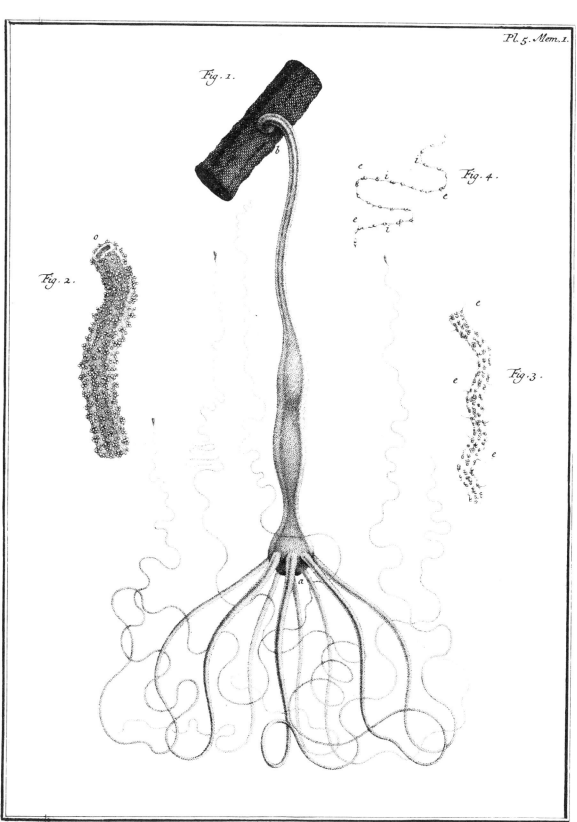

Fig. 1.

Fig. 2.

Fig. 4.

Fig. 3.

C. Pronk del. ad Viv. 1744.   J. v. Schley sculp.

MEMOIRS
CONCERNING THE NATURAL HISTORY
OF THE POLYPS.

❋ ❋ ❋ ❋ ❋ ❋ ❋ ❋ ❋ ❋ ❋ ❋

SECOND MEMOIR.

*On the feeding of polyps, on their manner of seizing
and swallowing prey, on the reason for the color
of the polyps, and some further discoveries about
their structure. On the season and methods best
suited for finding polyps.*

s indicated in the preceding *Memoir*, I
found the green polyps first. I observed
them for more than six months, but no
matter how carefully I attended to them, I
could not discover their mode of feeding. Once I had

become convinced that they were animals and knew a little about their structure, I suspected that the opening noticeable at the anterior extremity was their mouth. Before I could pursue my research any further, however, all my green polyps died. From April of 1741 until the present (January, 1744), I have sought in vain to find others.

In the middle of that April, I was compensated for the loss of my green polyps. I had stocked several jars with aquatic plants in the hope of finding some green polyps upon them, but instead I discovered on these plants some reddish polyps which were much larger. These were of the second species (Plate 1, Fig. 2). I soon satisfied myself that these polyps had the extraordinary properties which I had observed in the green ones, and I successfully performed experiments upon them that I would scarcely have risked undertaking on those of the first species because of their small size.

Shortly after discovering the second species of polyps, I learned how they were nourished. At that time the water in which I found them contained large numbers of a rather slender kind of *millepede* [a small freshwater oligochaete known today as *Stylaria*; see Book I, p. 48], only about seventeen millimeters long. They are unusual in that they have a fleshy proboscis or barb (Plate 6, Fig. 1, *d*) on the front of their head, a structure not found among other species of millepedes, and one which prompted Mr. Réaumur to call them *barbed millepedes*. They use very rapid inflections of their bodies to stay up in the water and to swim. They rest and crawl upon whatever objects they encounter and often are found in abundance on aquatic plants. The plants on which I first saw polyps of the second species were heavily bedecked with these barbed millepedes. Thus I drew the polyps and the millepedes together out of the water and without any special intent placed them in the same jars.

A few days later, as I was observing the anterior end of a polyp (Plate 6, Fig.2, *a*), I saw that a millepede had passed through this anterior end part way into the body of the polyp while the other part still protruded outside of it (Fig. 2, *m*). At first I did not know what to make of what I was seeing. I could not decide whether the polyp was devouring the millepede or whether the millepede had deliberately entered the gut of the polyp in order to feed off of it or to deposit eggs or its young there. Only reluctantly did I absent myself for a few hours from this spectacle which had so greatly excited my curiosity. Impatience to know what would become of the millepede drew me back to my study as soon as possible. No longer did I see the portion of the millepede (Fig. 2, *a m*) which at the time of my departure had protruded from the anterior end of the polyp, and I had reason to believe that it had passed into the body of the polyp. The body appeared swollen to me, and the transparency of its skin allowed me to see a small heap of matter inside that I had never before observed in the body of these animals. I judged that the millepede was dead and that moreover the polyp had partly digested it. In order to better satisfy myself that this was the actual case, I put some polyps with some millepedes aside in small shallow glass jars. From the first time I came to examine them, I found a polyp which was indeed swallowing a millepede. Since I was already accustomed to them, I recognized the millepede quite clearly then inside the body of the polyp. Thus I had hardly any further reason to doubt that the polyps were carnivorous animals.

Because I was extremely curious to learn how polyps seize their prey and how they carry it to their mouths, I hastened all the more to devise new experiments. I gathered more than a hundred polyps in a large powder jar and gave them time to disperse along the walls of the jar. At the moment when the majority had their bodies and arms extended, I threw in a large number of millepedes. They scattered at once, swimming all about in the jar. I followed them constantly with my eyes and before long I saw several strike against the arms of various polyps. As soon as they touched the arms, the millepedes were seized in such a way that their efforts to free themselves usually were useless. At first some were held by only one arm, others by several. To better secure the millepedes, the polyp then contracted and bent these arms and often enveloped the millepedes with still more arms. Finally, the arms drew them close to the anterior end which began to open, and through it the millepedes gradually were inserted into the body of the polyp.

The function of various parts of the polyps, unknown to me until then, was made very plain by the scene I have just described in general. I learned that those slender threads at the anterior extremity of the polyp, which I had already seen perform the function of legs, also served the polyps as arms. There was no further reason to doubt that the mouth was the opening the polyps have at the end from the edges of which the arms proceed, and that the stomach is the sac with which the mouth communicates and which extends to the other end of the body. I had already given the name "anterior" to the end of the body where the arms are located because I had noticed that when the polyps walked, this was the end that advanced first. When I saw that the mouth was located there, however, I had a new reason to call it the anterior end, and I did not hesitate to consider it the head of the polyp.

After discovering the polyps of the third species and observing the extra-ordinary length of their arms, I was also very curious to see them seize their prey and carry it to their mouth. They capture their prey in basically the same way as do polyps of the second species, but the length of their arms makes their maneuvers much more conspicuous. For this reason, I shall devote my attention principally to describing these maneuvers of the long-armed polyps. Once these are known, it will be quite easy to visualize those of the other species.

In order to see polyps of the third species seize their prey while their arms are greatly elongated, one must place the polyps in a glass seven or eight inches high. If the polyps are fastened at the top of the glass, their arms, for the most part, hang toward the bottom. It is most expedient to feed them while they are in this position in order to see their maneuvers in greater detail. Thus it is important to know how to get the polyps in place at the top of the glass.

One way to get them into place is to suspend the polyps from the surface of the water. This method, however, is not always convenient. Polyps that one raises are usually infested with tiny lice, and in order to rid the polyps of them, one should pass the end of a brush over the polyps' bodies several times in succession. If the polyps are holding only to the surface of the water, it is almost impossible to avoid making them let go. Thus, in this situation, I have adopted another expedient for suspending polyps, that of making them fasten onto a string (Plate 6, Fig.3, *f h b k g*). I submerge the area *b*, to which the polyps are fastened, slightly below the surface of the water and allow the ends of the string *h f* and *k g* to hang out on opposite sides of the glass. When the polyps are fixed on these little cords, one can pass the brush back and forth over their bodies repeatedly and even quite roughly without detaching them from the string. When one wishes to put the polyps in clean water, one has only to take the string by its ends, draw it gently out of the water and place it in another glass prepared in advance for this purpose. In order to have polyps attached to strings, one need only put a number of pieces into a jar stocked with polyps. There will always be some that will proceed to fasten themselves to the strings.

When the arms of such a polyp are well extended, I place a millepede or some other worm into the jar and, with the end of a brush, little by little I push it toward the tip of the arm by which I want it caught. It need but touch the arm for it to be seized. As soon as a millepede feels itself captured, it struggles vigorously, making great efforts to free itself. Often it sets out to swim (Plate 6, Fig. 3, *m c n*) and drags the arm *a c* by which it is caught from one side to the other as a fish caught on a hook drags the line if given slack. The first times that I witnessed this action, I expected at any moment to see the millepede, with its violent and repeated jerking, break the polyp's arm and then carry off a part of it. But experience has taught me that, however slender the arms of the polyp, they can resist considerable stress. I never saw any millepede succeed in freeing itself by breaking off an arm of a polyp. Rarely do they free themselves from the polyps at all.

The movements of the captured millepede eventually force the polyp to draw its arm back. At first the polyp partially contracts the arm, which then often shortens more as it twists near its point of origin in the process of contracting into a corkscrew shape (Fig. 3, *o i*). The millepede, continuing its struggle, entangles itself further in the arm that grasps it (Fig. 3, *m i n*). Frequently it hits other arms which, because of the jolting, are forced to contract and to draw near the head. Or the polyp itself may bring other arms near the prey to aid the arm that has seized it. In a moment, the millepede ends up entangled in most of these arms. By bending and continuing to contract, the arms soon carry the millepede to the polyp's mouth, against which they press it and hold it fast.

It is easy to imagine that swimming millepedes will encounter other parts of the polyps' arms more often than they do the tips. From one end to the other, the arms of the polyps are for the millepede what a limed-twig snare is for a bird. No matter what place on the arm a millepede may touch, it is seized and the nearer this is to the base of the arms the less difficulty the polyp has in bringing the millepede to its mouth. Midst all the arms, the prey then is entangled all the more quickly by its struggles. To bring the millepede near the anterior end of the polyp, the arms need but to contract and bend only those parts between their base and the place where the millepede is entangled. Often the part of the arm which extends from the spot where the millepede is entangled to the tip contracts only slightly and continues to hang down in the water while the other part is wound around the prey which it has seized and drawn to the mouth.

Up until now I have attempted only to depict the manner in which a millepede is seized and brought to the mouth of the polyp by arms which hang down towards the bottom of the water. The arms, of course, can be situated not only in this manner, but can extend in all directions (*Memoir* 1, p. 18), as is seen very frequently in observing the arms of a single polyp

(Plate 2, Fig. 3). When a millepede, or especially some larger and heavier worm, encounters a polyp's arm held nearly parallel to the bottom of the container, the weight of the captured prey often drags down the arm and even the body of the polyp, making them assume a position perpendicular to the bottom. The result turns out precisely the same as when the worm is taken by an arm already hanging perpendicular to the bottom of the vessel. At other times the polyp does not give its prey the time to drag down the arm which has seized it, but draws it back at once, entangles in its arms the prey which it is holding and carries it to its mouth.

This maneuver is performed most rapidly when the prey is grasped by an arm that is raised above the polyp's head and pointed toward the top of the container. The weight of the worm and the motion imparted to it when the polyp draws its arm back make it fall more rapidly than the arm can retract. Instead of being drawn in by the arm, it is the worm that drags the arm down, as a stone thrown in the air and then falling back would drag down a cord to which it was attached. The worm plunges down onto the other arms or onto the very head of the polyp where it is within immediate reach to be devoured. It sometimes happens that a polyp will use no more than one or two arms to carry the prey to its mouth and to hold it securely while swallowing it.

Even when a polyp has no prey at all to eat, its mouth is often open. The opening is ordinarily so small, however, that a magnifying glass is needed to find it. In contrast, as soon as the arms have brought a victim to the mouth, it immediately opens further and always in proportion to the size of the animal which the polyp is to take into its body. Little by little its lips spread out until they adjust precisely to the shape of the prey.

All the worms that polyps seize are not brought to the mouth in the same way. The manner in which the prey are delivered varies widely, depending on myriad circumstances that would be useless to specify because they can be noticed easily by all who study the polyps. In cases when one of the worm's extremities is presented to the mouth, the polyp does not need to open its mouth very much, only exactly enough to allow the extremity to enter (Plate 6, Fig. 2). Once the worm is grasped between the lips, the polyp extends the lips forward, and little by little the worm is made to pass into the body of the polyp. The arms do not appear to aid in the swallowing of the worm, and it is quite likely that it is the lips of the polyp which draw it in by a kind of suction. I am of this opinion because I often saw how polyps held millepedes or other animals between their lips with no arms touching the victims. I have seen a polyp whose arms I had cut off swallow a worm with as much ease as did those with arms. The victim certainly was not making any efforts to enter the stomach of its enemy. On the contrary, it is quite reasonable to conclude that all its energetic movements were signs that it was resisting.

If the ingested worm is no longer than the stomach of the polyp, it often remains stretched out inside. If it is longer, however, the end which entered first bends around so that when the worm is completely swallowed it lies folded double in the stomach (Plate 6, Fig. 5).

When it is not the end of the worm which is presented to the mouth, but rather the middle, or some other part of its body, the polyp grasps that part with its lips, extending them to the right and the left and applying them against the worm (Fig. 4). The mouth then takes the form of a boat pointed at both ends. After that the polyp draws the two points of this boat a little closer together. This movement and the suction compel the worm to bend and fold in two, and it is swallowed in this position (Fig. 6).

The skin of the polyp stretches proportionally as the stomach fills. The volume of the stomach increases, and the body becomes shorter, wider and more thickset (Fig. 9). When it is full, the polyp hangs motionless as though in a kind of torpor with its arms usually quite contracted. Its shape then is far different from when it was extended (Plate 7, Fig. 6). As digestion proceeds and as the polyp voids what is not useful for its nourishment, the body shrinks and elongates, little by little resuming its original shape. When digestion is completed, the polyp ordinarily appears elongated again, and often is ready soon thereafter to seize the next victim.

For the present, I shall not go into greater detail about the manner by which a polyp swallows its prey, or the condition of the polyp's body after it has swallowed it. I shall soon have the opportunity to present certain facts which will lead me back to these matters.

Having seen polyps eat millepedes and having ascertained that they constituted a most suitable nutriment for the polyps, I collected a great quantity of them in order to have the wherewithal to feed the large number of polyps which I was keeping in glass jars. My task of gathering these millepedes provided me with the opportunity of observing them in the ditches from which I took them. I saw certain locations in these ditches which swarmed with millepedes. They were crawling on the plants and on all the other objects in the water, swimming around from one plant to another.

Then I easily surmised how polyps get within reach of their food. I understood that it was for this purpose that the polyps climb upon plants and disperse over their branches. Indeed, as soon as they reach an appropriate location, they need only extend their arms and wait until those arms are struck by any of the millepedes swimming to and fro in the water. When fully extended, the arms of a polyp cover a considerable area, one which a millepede traverses only at great risk of being captured. If a polyp of the third species extends its arms at some length on all sides around its head, the snare it thus sets for its prey can reach a foot in diameter. The polyp then resembles a spider that sits in the middle of its web, waiting for some small fly to fall into the trap. Then again, the polyp could be compared to a fisherman with a line, but this would be a fisherman who uses several lines at once. While some of the arms are engaged in holding one victim and bringing it to the mouth (Plate 6, Fig. 3, *m i n*), the others often remain outstretched, seizing whatever additional prey presents itself (*m c n* and *p*).

Frequently a number of millepedes are seized at almost the same time by various arms of a single polyp (Fig. 3), which then devours them one after the other.

When the polyps extend their arms, they do not have designs on the millepedes alone. These traps are set equally for most of the small creatures that swim in the water. The first to come along are the first taken.

From the time I noted the voracity with which the polyps ate the millepedes, I conjectured that they were not the only suitable prey for the polyps. I hastened all the more to learn which other animals could serve as food for them, because I was having difficulty obtaining sufficient millepedes to provide ample sustenance for the polyps I was raising. Besides, collecting the millepedes required a great deal of time.

I cut open some satiated polyps that I had taken from the ditch and removed from inside them some small water fleas [*Daphnia;* see Book I, p. 48] which were still quite recognizable. Among these was a species that multiplies prodigiously and can often be obtained easily in great quantities. The one depicted in Plate 6, Figure 3 at *p* is shown actual size, but a strong magnifying glass or a microscope must be used to discern the details of its appearance (Fig. 11). Swammerdam gives an exact description of these insects (Leiden edition, 1737, p. 86 and following). Their most noticeable feature consists of two branching arms rising above their heads (Fig. 11, *b, b*) which are used as fins. Diverse movements, too numerous to describe, are made with these arms. The shape of their arms led Swammerdam to give them the name of water fleas with branched horns. These fleas, usually reddish in color, jump about constantly in the water.

When I placed some of the water fleas I had collected into a powder jar containing polyps, the polyps quickly seized them, even as I watched. One could say that when they touch an arm of a polyp, the water fleas know at once the danger they face. Within an instant they begin to struggle vigorously. If a water flea is caught at or near the tip of the polyp's arm, it often proceeds to drag the arm along (Fig. 3, p), as I have said the millepedes do. Frequently I have seen water fleas escape successfully. It seemed to me that they set themselves free more easily than the millepedes. Because the water fleas are quite small, and especially because their bodies are not elongated, when they struggle they run less risk of entangling themselves in the arms of the polyp than do the worms.

At times the polyp will use but one arm to seize a water flea and carry it to its mouth. In this case, it ordinarily shortens the arm by twisting it into the shape of a corkscrew. It should be noted that the polyp does not twist the arm about all the way up to its base. A rather thick portion about five millimeters long always remains uncoiled. This is the portion which bends back against the mouth, bringing to it the part of the arm holding the water flea.

The water flea is a little longer than it is wide (Plate 6, Fig. 3, p and Fig. 11). If this animal were held against the anterior end of a polyp at a time when the polyp's mouth is not devouring any prey, it would be as though a man were holding to his mouth a pear as big as his head. A man would manage to ingest this pear by breaking it into pieces with his teeth. But the mouth of a polyp has no teeth or hard parts to perform this function. The polyp makes up for the absence of teeth by its ability to open its mouth so prodigiously wide that it can swallow the water flea whole regardless of the position of the flea when it is brought to the mouth. Soon after the victim has been pressed against the anterior end of the polyp, the mouth widens and gradually forms a hollow, like a kind of cup, into which half of the water flea is packed (Fig. 7, a). As the lips of the polyp continue to extend, they stretch next over the half which is still outside the mouth (Fig. 7, p). Little by little this portion is completely covered by the lips as they come together, and once again the mouth of the polyp closes. At that time the water flea is distinctly visible through the skin of the polyp below the mouth. Since the body of the flea is wider than that of the polyp, the polyp is forced to bulge in the spot where the flea lodges. This first water flea is often followed by another which drives the first further down into the stomach. A third in turn pushes the second, and in this way, depending on the size of the polyp, four or five water fleas in a row can be seen being passed into its body, one pushing against the next. Once they are all inside, the body of the polyp appears to be almost uniformly thick throughout, conforming to the widths of the water fleas within it. If there is some irregularity in the thickness of the polyp's body, it occurs because all the water fleas did not pass into the stomach in exactly the same position, some entering widthwise and others lengthwise. Thus the thickness of the polyp varies according to the difference between the width and length of the ingested water fleas. Moreover, the points at both ends of these little animals (Plate 6, Fig. 11, *c e*), which are noticeable under the skin of the polyps, force the skin covering them to jut out in points. Because the whole exterior of the water flea is horny, the skin of the polyp, characterized by great softness and flexibility, is forced to yield to the hardness of the extremities of the water fleas.

The polyp does not stop after having swallowed four or five water fleas. If it is holding others in its arms, it continues to swallow them. The earlier victims, having to make place for the latter, can no longer remain in a single file. Thus, the water fleas are crowded together once again, forcing the walls of the stomach to expand in width until often its capacity is sufficiently enlarged to allow two water fleas to be set widthwise alongside each other (Plate 6, Fig. 8), though quite unsymmetrically. When it is extremely hungry, a good-sized polyp can easily swallow a dozen water fleas. When all these fleas are in the body of a polyp of the second species, it is absolutely full from the mouth to the posterior tip (Fig. 8). In a polyp of the third species, the portion of the body which narrows down into a tail usually remains empty and tapered (Fig. 9, *b d*). At times even this portion is forced to expand to receive several water fleas. When the polyp has not devoured as many water fleas as its stomach can hold, its body is often quite slender near the head, forming a very conspicuous neck there (Figs. 9 and 10, *c*).

Because I had a considerable number of polyps to maintain during a span of more than two years, I needed to supply myself with a store of food appropriate to their needs. The water fleas with branched horns were a very great resource for me during several months of the year. I believe I can safely assert that no other creatures furnish as plentiful a prey for the polyps as do these water fleas. They are seen in the greatest abundance on warm days when the weather is calm. At such times, large areas of ditches that are full of them are tinted with a reddish color. I have often seen these bands about a foot wide and more than fifty feet in length; there the water fleas were so dense that absolutely nothing else was visible. They seemed to concentrate particularly in areas where the sun shone. Perhaps the sunlight attracts them. At least candlelight certainly attracts them, as the following experiment persuaded me:

By candlelight, I was observing some polyps to which I had fed water fleas during the day. That evening a group of fleas which the polyps had not eaten remained in the jar. I noticed that most of these water fleas were assembled at the side closest to the candle. I changed the location of the candle; the water fleas also changed theirs and came to the side where I had put the candle. After changing its location many times and seeing that the water fleas drew near it without fail, I rotated the candle around the jar, rather slowly indeed, but without stopping. The water fleas followed it, circling the jar a number of times. I repeated this experiment on many occasions.

If a large number of water fleas are suddenly thrown into a jar containing hungry polyps, the arms of the polyps are soon laden with them. I have seen arms several inches long that were so loaded down with water fleas from one end to the other that they were completely hidden under them. The activity of these little animals trying to escape forces the arms to contract. Soon nothing is visible but a jumbled mass of water fleas clumped near the anterior end of the polyp (Plate 7, Fig. 1, *a*). The polyp then swallows them one after the other until it is entirely full. If afterwards there are fleas still attached to the arms of the polyp, they often remain there for a time. Sometimes such fleas die, and at other times they escape.

If a polyp has not taken enough water fleas to satisfy it at the outset, its arms remain partly elongated, still prepared to seize prey. In this way the arms snatch and carry to the mouth those water fleas that present themselves, and the polyp swallows them. Thus, provided it does not develop an aversion to food, a polyp situated in an area where it is always within reach of prey will never go unfed. I often observe polyps which stop eating for a while and fail to grasp available prey even though their stomachs are empty. At times this aversion to food marks the start of a fatal sickness; at other times a return of appetite ensues.

As long as I was able to procure an abundance of water fleas, I preferred them to all other food for my polyps. No other is simpler to collect. For this purpose I use a kind of hoopnet made of a brass wire circle eight to ten inches in diameter to which I attach a pouch of thin cloth. I fit this hoopnet to the end of a stick and place it under water in an area well stocked with water fleas. Holding the hoop slightly at an angle, I move it about in the water wherever I can reach. Once the water fleas are gathered in the hoopnet, I draw them out of the ditch and place them in a small amount of water which soon swarms all over with the little creatures. Finally, I proceed to put a few drops of the water into each of my jars and, with these drops, hundreds of water fleas. The entire process is completed in less than a quarter of an hour.

From the beginning of June, 1741 until the end of the month of September, I always had as many water fleas as I needed to feed the large number of polyps on which I was experimenting. When the water fleas began to be in short supply, I had some difficulty at first in finding something to feed my polyps. Several times a day I would go to the edge of that ditch which for so long a time had furnished me with water fleas. I would bend down very close to the surface trying to discern some young ones on which to build my hopes for the future. My searches were to no avail. In searching for water fleas, however, I discovered places at the bottom of the ditch absolutely bristling with worms [*Tubifex;* see Book I, p. 48]. One end of each worm was in the earth, whereas the rest of its body projected above and undulated incessantly in the water (Plate 7, Fig. 2). From the moment I saw these worms, I thought it likely that they would be useful as food for my polyps. I hoped they could compensate for the dearth of water fleas and for the millepedes which I could then find only in very small quantities and with a great deal of difficulty. I took a few of these worms and gave them to some polyps, and the polyps ate them.

Then I had to find a way to obtain as many worms as I would need. Although they were abundant at the bottom of that ditch, the difficulty lay in collecting them. As soon as I tried to get hold of them, they would withdraw entirely under the earth. I followed the expedient of putting some of this earth in pans and looking for the worms there. This method proved long and tedious, however. Finally I found an extremely quick and convenient technique.

I attach a hoop of iron wire two to three inches in diameter to the end of a stick. I place it in the water and insert a segment of the hoop underground to a depth of one or two inches. Holding the instrument steady in the same position, I draw it along over a small area. The iron wire encounters the worms underground and drags them along until it is pulled out laden with them. Then I immerse the iron wire hoop in a jar full of water and shake it so that all the worms fall to the bottom.

In order to catch worms with this instrument, the bottom of the ditch must be quite clean. If it is covered with leaves and grass, the debris gets around the iron wire so that it has little or no grip on the worms. As a result, it catches few worms, or often none at all, although they may be plentiful in the area where one is fishing for them. In certain locations a few strokes of a rake are sufficient to clear the area. There are other areas, however, where it is not possible to remove everything which could get around the wire and prevent it from catching the worms without carrying away the worms themselves. In that circumstance, it is more expedient to spread a few inches of sand over the littered bottom. Since the worms must stay near the surface of the earth in order to project a part of their bodies above it, they leave the mud, pass into the sand, and come to this new surface. Twenty-four hours after treating the bottom of the ditch in this manner, I have been able to capture great quantities of worms.

To feed the worms to the polyps, one must carefully drop them on the arms of the polyps. Unlike the water fleas and the millepedes which can be caught as they swim by, the worms fall to the bottom of the jar as soon as they are put into the water. It would be useless to describe at length how the polyps seize these worms, carry them to the mouth, and swallow them. To execute these maneuvers, the polyps employ the same methods they use with the millepedes and water fleas. Thus, I shall mention only a few distinctive aspects of the manner in which they feed on the worms.

An elongated polyp can eat a worm much longer than itself and at least as thick as its own body. By dint of stretching its own skin and folding the worm over and over a number of times, the polyp finds a way of packing it into its stomach. There are no hard parts in the worm, nothing to offer resistance to the skin of the polyp. It yields easily and lodges inside the stomach (Plate 6, Fig. 5). Even before being inserted into the stomach, the worm often has already been rolled and folded into a bundle through its struggles amidst the arms of the polyp which is trying to subdue it. In order to accommodate a worm bundled up this way, the mouth of a polyp must open extremely wide. One can often see a worm that has been folded three or four times passing into a polyp's stomach. When it is simply folded in half at about the middle, the two ends of the worm are clearly visible still hanging outside of the mouth while the central portion is being pulled inside the body of the polyp (Plate 6, Fig. 6).

These worms constituted one of my best supplies of food for the polyps, especially during winter. I had collected an abundance of them in October and placed them in large vessels full of water with three or four inches of earth at the bottom. Whenever I needed them, I fished them out of these vessels in the same manner as I had fished them from the ditches.

It must be evident that in order to find the worms I have described, one must probe the earth at the bottom of the ditch with the iron wire instrument used in catching them. With this instrument one also often gathers thick red worms about twelve millimeters long (Plate 7, Fig. 8, *c d*) from under the earth. This worm [larva of *Chironomus;* see Book I, p. 48] belongs to the same class as the one described in the first *Memoir* of the fifth Volume, page 29 and following, of the *Memoirs* of Mr. Réaumur on the insects.

The polyps may be fed this worm as well, but it is more difficult for them to swallow and to digest than those discussed previously. Because the worm is thick and rigid, it is not easy for the polyp to make it fold over and lodge in its stomach. When a worm that is folded in two is inserted, it fills up a very large area inside the polyp and forces the skin to stretch a great deal. The skin of the worm is somewhat scaly, making it more difficult to digest. The polyps have to be quite hungry before they will eat it. It is not suitable food for them during winter.

I have seen polyps eat the transparent larva of the gnat [*Chaoborus;* see Book I, p. 48] that Mr. Réaumur describes, starting on page 40 of the *Memoir* cited above.

When in June, 1743 I caught a large number of little fish about nine millimeters long, my first recourse was to test whether or not polyps would eat them.

I put several of these fishes into jars containing many polyps. The experiment soon proved to me what I had suspected from the start: the liveliness and strength of these little fish enabled them to resist vigorously. I was none too hopeful that the polyps would manage to seize them. As they swam about, these roach fry, for this is the species of fish in question, soon encountered the arms of the polyp and then the battles began. Not all ended in the same way. If the fish met but a single arm, it was usually able to escape by jerking sharply, quite often even breaking off the arm that held it and carrying away part of it. For the fish seized by several arms at once, however, the battle ended less fortunately. Most of the time the efforts it made to free itself were not only useless, but even tended to entwine it all the more in the arms of its enemy. It was obvious that the polyp was exerting itself strenuously to hold on to the fish. The arms which enveloped the fish on all sides and which were fastened tightly around it were quite swollen, a condition that rarely occurs except when they are straining forcefully. In a word, what Ovid says of the marine polyp may be applied perfectly to the freshwater polyp described here. The poet could be said to be speaking of that latter animal when he writes (*Metamorphosis,* Book 4):

> *And so beneath the sea, on all its sides the polyp*
> *Stretches its tentacles to seize and clasp its prey.*

When I saw a polyp that had seized a fish and that was drawing it to its mouth, I thoroughly expected it to do everything possible to swallow it. It was a matter of the polyp passing into its body a rather thick fish nine millimeters long which could not be folded to lodge in the stomach. The polyp that was attempting to swallow the fish then measured hardly more than five to seven millimeters in length, having been forced to contract to that size by the joltings of the struggling fish. Nonetheless, most of the polyps that seized a roach fry managed to swallow it. When a polyp with long arms was swallowing a fish, the narrow portion of its stomach forming its tail was forced to open and take in part of the prey. A polyp which had swallowed a fish was difficult to recognize. Suppose, for example, that the fish had been swallowed tail first; then one would see contracted arms at the tip of the fish's head (Plate 7, Fig. 3, *a c, a c, a c*), and that is what the creature most resembled. The skin of the polyp *a b* was so perfectly stretched and fastened against the body of the roach fry that the fish showed plainly through it (Fig. 3). Often someone not otherwise informed might believe that he was seeing simply a fish with barbels about five millimeters long at its front end.

This fish, packed completely whole inside the body of a polyp and stretching the skin of the polyp very thin, was nevertheless digested there. It did not remain alive a quarter of an hour. Later, after the juices had been sucked out of the fish, its remains were regurgitated from the mouth of the polyp, recognizable it is true, but even so quite disfigured. I have seen this process take place a great many times.

Polyps will eat most small creatures found in fresh water. They feed very well on larvas and pupas of gnats and of other small flies. Lastly, the polyps can be fed much larger animals provided those animals have been cut into small pieces. In this way I have fed them slugs and other still larger aquatic creatures, as well as earthworms and the entrails of freshwater fish. What is more, for a time I fed them with meat from the butcher's shop: beef, mutton, and veal. I saw plainly that they drew nourishment from it, observing them grow during that period. It is true, however, that they did not obtain as much nourishment from this meat as from the aquatic creatures which suit them best. In order to make them eat the meat, one must cut it into very small pieces because it swells once it is in the water.

The most important task in raising polyps is to find suitable food for them during all the seasons of the year. In the summer, so long as there are ponds or ditches nearby, it is easy to procure some of the prey I have discussed so far, but it is more difficult to find them in the winter. Then my principal supply consisted of those worms which ordinarily stay underground at the bottom of the water, and which I maintained in jars (Plate 7, Fig. 2). Fearing that I might run short of them, I sought other expedients to feed my polyps. I placed into the bottom of a jar some earth taken from a ditch, reckoning that with this earth I would have deposited a good number of small creatures, or at least the eggs from which they would emerge. This strategem succeeded very well for me. From the end of February, 1742, my jar teemed with various species of small animals. Little by little, however, it became particularly well-stocked with a kind of small creature enclosed in a bivalve shell [an ostracod; see Book I, p. 48]. In order to swim, it opens its shell part way and extends little legs or arms which it moves about very rapidly. This animal, about the size of a grain of sand, alights on anything it happens upon. I put polyps into the jar with them, and I paid them no further attention. The polyps fed on these animals and multiplied. I have some polyps that for eight months have been in such a jar, the bottom of which is covered with earth taken from the ditch. They are feeding there, and they are multiplying.

During the summer months, a very suitable habitat can be arranged for the polyps by preparing some tubs with their bottoms covered by earth taken from ditches. If the tubs are exposed to the air, in addition to the creatures from the eggs in that earth, others soon will be seen emerging from eggs that have been deposited on the water by crane flies and gnats. In this way, these tubs and jars can be used successfully both to spare oneself the trouble of feeding the polyps kept in them and to prevent a scarcity of food which can disrupt interesting observations.

Often it is convenient not to change the water of these jars in which the creatures are kept. When it is not changed and the vessel is not cleaned, however, it frequently becomes filled with a grass, fine as hair, in which the polyps become entangled. Or alternatively, minute plants that are almost imperceptible grow upon the walls of the container, so obscuring them that if the receptacle is made of glass, it becomes impossible to see inside it distinctly. There is an easy way to prevent these inconveniences. Place some aquatic snails in each vessel, the number of snails depending on the size of the container. The snails will eat these plants as they grow, and the water and the sides of the container thus will remain clear.

I stated earlier that a creature had merely to touch the arm of a polyp for it to be seized. I have had frequent opportunities to observe this. I even saw a millepede which touched an arm with but a single one of its feet. It struggled briskly for some seconds, without being able to detach itself.

Whatever the cause of this phenomenon, it is not at all automatic but is under the control of the polyp. To become convinced that they have such control, one has only to drop on the arms of thoroughly satiated polyps some animals they like very much to eat. Frequently, the polyps allow the prey to slide over their arms without restraining them; had the polyps been hungry they would have seized them the moment they touched the arms.

When a hungry polyp is given things which cannot serve as its food it will sometimes first grasp them with its arms and then let them drop. At other times it will not take hold of them at all.

If the arms of the polyps were just like limed twigs, if they were coated with a glue which caught hold of all the objects that touched them, and if the arms were not capable of stopping this action, one would expect two events to occur: instead of seizing only some of the time the objects they encounter they would seize those objects consistently. In addition, when these arms touched each other they would of necessity fasten together very firmly. It is easy to observe, however, that the arms touch and even wind about each other without becoming stuck together. At times, I have enjoyed picking a moment when all the arms of a polyp of the third species were well extended; then I would entangle them all together and force them into a bundle that looked more difficult to unravel than a badly jumbled skein of yarn. At first I thought that the polyp would never manage to untie this Gordian knot and that I would need to cut it in order to set the arms free. Yet the polyps always succeeded in disentangling themselves, although some required two or three days to do so.

Thus, from all that I have said, it would appear that it is within the polyp's power either to set in motion, or not to set in motion, whatever operations are needed to clasp the animals that touch its arms. I can say no more on the subject with any accuracy. Perhaps the cause governing this action is the same as that which leads the arms and body of polyps to adhere to the objects with which they come into contact.

At first one may be inclined to compare the beads so readily noticeable on the arms of polyps, especially when they are extended (Plate 5, Figs. 2, 3, and 4), to the knobs, that is the hollow rounded protuberances that line the feet of marine polyps and the arms of cuttlefish. They are described in the works of the authors who discuss these animals, especially in the account of the arms of the cuttlefish by Swammerdam (*Bibl. Nat.,* p. 877 and following). It is clear that in cuttlefish these knobs serve to fasten the arms against the objects they touch, and it is quite obvious, as Swammerdam thought, that this adhesion is accomplished by a kind of suction. I have pressed the arm of a dying cuttlefish against my hand and felt its protuberances still clinging forcefully to my skin.

The structure of the beads on the arms of freshwater polyps did not seem to me to bear any similarity to the structure of the protuberances on the arms of the cuttlefish and the marine polyp. In the first *Memoir,* I showed that these beads were formed by the clustering of some of those granules with which the arms and bodies of the polyps are filled. It is possible that the polyp effects a kind of suction which serves to hold the prey. It may do so by firmly applying the edges of the part of its arm with which it touches the prey, while retracting the middle portion of that part of the arm. I have often noticed that when a creature was resisting vigorously, the section of the polyp's arm holding it fast would swell considerably.

Animals which can serve as food for the polyps are not evenly distributed throughout a ditch. Do polyps know how to choose the best stocked areas, or do they wander here and there guided by chance alone? All I can say on this subject is that their propensity for the best-lit areas may lead them to places well stocked with animals suitable as food for them. The facts I reported in the foregoing prove at least that the water fleas with branched horns, which constitute one of the principal foods of the polyps, seek light and also proceed to the best-lit areas.

Once the polyps have settled in a certain spot, however, do they have no other recourse but to extend their arms and wait until an animal comes along and encounters them? Or do they have a certain sense which enables them to perceive their prey, and when they have perceived it, direct their arms towards it so as to seize it? I have performed some experiments which perhaps could provide some insight into these questions.

An incident recurred a number of times while I was feeding one polyp four to five millimeters away from another. The prey did not in any way touch the arms of the second polyp, which were turned in another direction. Nevertheless, the second polyp immediately bent its body around and brought several arms toward the prey.

One day I wanted to feed a young polyp which was still attached to its mother. Choosing a moment when the heads and arms of each were turned in opposite directions, I dropped a small worm on the arms of the young polyp. Instantly, the mother turned its head and set about seizing the worm. Often in such a case I allowed the prey to be taken by the mother. This time, however, I took the worm out of the water. Then I sliced off the mother's arms completely, cut off her head, and gave the worm back to the young polyp, assuming that the prey could no longer be stolen from it. Once the young polyp had seized it, I stopped observing it, and only returned to look at it again about half an hour later. At that moment I saw something I was not anticipating. The mother's lips were turned back on the outside of what remained of her body; and the worm, which I was expecting to find inside the stomach of the young one, was in the process of entering the stomach of the mother. It was going in not through what could be called her mouth, but rather through the opening that was formed at the end of this headless stump by the turned-back edges of the anterior portion of the mutilated animal. I did not stop watching until I had seen the worm swallowed completely.

The facts I have just reported made me suspect that the polyps had some sense which enabled them to perceive their prey. Thereafter I was more attentive in noticing anything which might confirm or discredit this suspicion.

At various times I have seen polyps in my powder jars that were fastened to branches of horsetail plants [most likely water-milfoil, *Myriophyllum;* see Book I, p. 48], turn their arms toward millepedes which were crawling on these plants and seize them.

At the bottom of a large jar I placed a worm which could not swim; attached at the top of that jar were some long-armed polyps. I even arranged it so that the worm could not budge from the spot where I had put it. My intent was to see whether or not the polyps which were situated some five or six inches away would come seeking it with their arms. They often did so.

When observing an animal, it is natural enough to investigate whether or not it has eyes. But it is especially natural to look for eyes in an animal which shows a marked propensity for light. Accordingly I did not neglect any opportunity to ascertain whether or not polyps had eyes. Although I have observed all areas of their bodies carefully with the magnifying glass and the microscope, I have never succeeded in discovering any part which, by its location or by its structure, gave me reason to suspect that it was an eye.

I have been unable to discover any eyes at all in polyps, and even the most skillful observers, aided by the best microscopes, may be unable to discover any. Yet it would be rash, it seems to me, to assert that they absolutely have none. It would seem to be more rash still to come to a general conclusion that polyps have no means of perceiving light or the objects it renders visible. When facts are lacking in such research, it is more appropriate to suspend judgment rather than to make decisions which almost always are based on the presumption that Nature is as limited as the faculties of those who study her.

Some of the observations recounted above will already have suggested that two or more polyps seizing the same worm fight over it. Two of them often can be seen, for example, tugging energetically at the same worm. Quite frequently it happens that one polyp begins to swallow the worm at one end and the other polyp at the other end, both continuing to swallow from opposite directions until their mouths touch (Plate 7, Fig. 4, *a*). Sometimes they will remain pressed against each other for quite a long time, after which the worm breaks and each polyp has half. At other times, however, the contest does not end there, and the polyps continue to wrangle over the prey even when their heads touch. Then one of the polyps opens its mouth wider and sets about swallowing the other along with the portion of worm it has inside its body. In fact it more or less ingests the other polyp, frequently almost in its entirety (Fig. 5). This contest, however, ends more fortunately for the polyp swallowed by its adversary than the observer would at first be led to suppose; its only loss is the prey which the other often wrests from its stomach. The polyp itself emerges from the body of its enemy completely intact and safe and sound, even after having been there for more than an hour.

Very frequently polyps, especially those of the third species, swallow a few of their arms in part along with their prey. The portion of the arm remaining outside often forms a readily noticeable loop (Plate 6, Fig. 5, *e, i*). Recall that in seizing its prey, the polyp wraps its arms around it. Instead of disengaging the arms as the prey is swallowed, the polyp leaves them where they were. Thus, when the worm or water flea is inside the stomach, it is still as entwined in the arms as it was before it was ingested (Fig. 5). What is more, the ends of these arms often remain in the stomach for over twenty-four hours, but at the end of that period they emerge just exactly as they were when ingested. Much less time is required inside the stomach, however, to macerate and digest prey having parts a great deal more solid than the arms of the polyp.

These events, especially the fact reported above of the polyp having emerged safe and sound from an hour-long stay within the body of another polyp, made me conjecture that these animals were not a suitable food for their own species. Since beginning to raise them, I have had an opportunity to ascertain in another way that such is the case. I had always kept a large number of polyps in jars where they were within reach to eat each other. In fact, at the start of my observations, I feared this possibility. While observing other animals, more than once I had seen those of the same species eat one another and by so doing deprive me of the pleasure of observing them for a longer time. I was soon reassured about the polyps, however. Although I have left them without food for days, and later for entire months, I have never seen one attempt to eat another. I even tried to induce them to do it. I took a polyp and presented it to another, just as if it had been a worm. Instead of seizing it at once, as it would have seized a suitable prey, the polyp made not the slightest movement to do so. Instead, it allowed the proferred polyp to slide over its arms and fall to the bottom of the jar; or if it did stay fixed to the arms, it was not for long.

Later I shall describe how I managed to introduce a polyp into the stomach of another polyp and force it to remain there [*Memoir* IV, p. 169]. For the present, I will say only that none of the polyps subjected to that experiment died even though some remained in the stomach four or five whole days. And it is important to note that among all the animals I have discussed as constituting food for the polyps, I have not found a single one able to survive in their stomachs for more than a quarter of an hour. When I forced a polyp to regurgitate a prey immediately after the polyp had swallowed it, I often drew it out alive. If, on the other hand, I tarried ever so short a time I always found it dead.

Having nurtured polyps of the second and third species through all the seasons of the year, I have learned that there is no season during which they do not eat except in the wintertime when the water temperature drops very close to the freezing point. The cold numbs them and suppresses the activity required for them to seek food and seize the prey they encounter. It causes them to lose their appetites completely, thus rendering all food superfluous. In these circumstances, when a worm is dropped on their arms, they do not appear to have any desire for it and do not even grasp it.

As the weather becomes warmer, their appetites revive and with it the strength to execute the maneuvers necessary for capturing prey. At the same time most of the creatures that the polyps use for food are either recovering from the state of torpor to which the cold had reduced them, or are hatching from the eggs that had been laid previously at the bottom of the ponds. These animals begin to appear and to expose themselves in their comings and goings to the snares that the polyps spread out for them.

It is not possible to correlate precisely an increase in warmth with an increase in the appetites of the polyps. A few degrees of warmth more or less do not produce a sufficiently perceptible change. Suffice it to say that the appetite of the polyps is much greater during summer when their voracity is indeed most remarkable. A polyp swallowing a worm at least as thick as its own extended body and three or four times as long (Plate 6, Fig. 4) is a common sight. I have already mentioned that in a single meal a polyp can eat ten or so water fleas or indeed three or four millepedes. In comparative terms, the volume of food which a polyp can consume at a single feeding is three or four times greater than the volume of its own body.

During the summer, no matter how great the quantity of food consumed in a single meal, the polyps digest it much more rapidly and are ready to start on a new meal much sooner than in any other season.

When the weather is warm, digestion is often completed at the end of twelve hours: the polyp has already extracted the nutritive substance from the animal it has swallowed and has discharged the wastes, emptying its stomach, which formerly had been filled with a great volume of material. Its body and arms are extended anew. For all this to be accomplished during a colder season, even though the polyp then eats far less than in summer, it often takes two or three full days, depending on the degree of cold.

Polyps discharge their wastes through the mouth. I have never seen anything come out through the opening which they have at their posterior end.

As is the case with most voracious animals, if polyps can eat a great deal at one time, they can also go quite a long time without eating. The life histories of insects provide us with examples of bees, ants, various species of caterpillars, worms, butterflies, and flies which can go through entire months without eating anything whatsoever. This period of fasting for these insects, however, is also a time of inactivity and of the torpor brought on by the cold weather. They could not remain without food for an equal length of time during the summer when they are active, albeit even then they are able to endure fasts of much longer duration than so many of the quadrupeds and other animals known to us. By contrast, even in the warmest weather the polyps can dispense with food for a very long time without dying. At such warm temperatures I have kept polyps in jars without any food for four months.

I made a careful effort to gain some insights into how the polyps digest their food and extract the nutritious material from it, and how this substance passes into their bodies to nourish them. I have never flattered myself that I had acquired very precise ideas on the subject. The following is all that I have been able to discover.

After the millepedes and the worms which I often feed to the polyps are introduced into the stomach, they are at first quite recognizable. The transparency of the skin of the polyps permits the ingested prey to be seen distinctly, most especially if the polyps have not eaten for some time (Plate 6, Fig. 5). By continuously observing worms in the polyps' stomachs, one can see that little by little they lose their shape and finally are no longer recognizable. The substance of which they were composed is reduced to a pulp containing various sized fragments of the more solid parts of these animals. All these observations lead one to believe that the foods are first macerated in the stomach of the polyps; then, after the nutritious substances are separated out, the remainder is discharged through the mouth.

When maintaining numerous polyps, one often has an opportunity to see excrements coming out through the mouth. These consist of parts of matter large enough to be easily discerned. If a polyp has not digested a worm completely, as is common in temperate or cold weather, one can find among the wastes some rather long pieces of worm, which by then are whitish in color and parts of which seem to have been only slightly torn (Plate 7, Fig. 6, e).

After the polyp has thrown off these wastes, its body is not always entirely free of swelling. Sometimes it appears full of a fluid which undoubtedly contains the juice that has been extracted from the food. This fluid still contains waste materials, but they are more difficult to observe coming out than in the earlier discharge I mentioned. Or, rather this fluid itself is nothing but the waste that remains after the parts destined for nutrition have been separated and have passed into the skin. This fluid comes out of the stomach in extremely fine filaments, as fine as the arms of the third species of polyp. Seeing this substance emerge from the mouth, I mistook it at first glance for the arms of these polyps.

The body of the water flea is not broken down in the polyp's stomach as is the body of the worm I have just discussed, the reason being that the worm's body is composed only of soft parts, whereas the entire surface of the water flea is of a horny material. These water fleas are usually reddish and often quite vividly colored. By observing them under a magnifying glass it is easy to determine that whatever gives them this color is in the interior of their bodies and not on the surface. The horny covering is white and transparent, revealing the viscera within that contain the colored substance which makes the water fleas appear red. The only visible change that can be discerned in a water flea between the time that it enters the stomach of a polyp to serve as food for it and the time it comes out as excrement is in its color; it enters reddish and it comes out white. Thus it appears that the reddish substance simply has been drawn out of the viscera of the water flea, presumably by a kind of suction.

My observations while feeding another kind of animal (Plate 7, Fig. 7) to the polyps seemed to confirm the above idea. This animal [*Glossiphonia;* see Book I, p. 48] is rather common. It moves by successively bending and extending its body. The body is a transparent white and has more or less the consistency of wax. The viscera, which are distinctly visible (Fig. 7) in most of these creatures, are full of a substance that is a beautiful crimson red. The one depicted in Figure 7 is too large to be eaten by a polyp; one must select for that purpose those that are only half the size. These animals also emerge from the body of the polyp without having been macerated, with only the red substance in their viscera having been drawn out. I opened up a polyp about an hour after it had swallowed one of these creatures, and I found that the red substance was in the process of coming out at one end of the worm. This material did not appear to be liquid, but seemed rather to have the consistency as well as the color of currant jelly. How is this red substance drawn out of the body of these worms? Would it not be by suction?

The following experiment suggests the same conclusion as the two preceding observations. I gave a red larva of the midge [*Chironomus;* see Book I, p. 48], discussed previously on pages 63-64, to a rather small polyp. The larva was very large and the midsection of its body was proffered to the mouth of the polyp. Consequently the polyp could not swallow this worm except by opening its mouth extremely wide. The first time I came to observe the polyp subsequent to the feeding, I found its lips fastened tightly against the worm, taking in a portion of the worm's body some five millimeters long (Plate 7, Fig. 8, *e i*). Thus the polyp's mouth had been forced to open considerably. Its body *b e i* took on the shape of a funnel flattened on two sides. I would not have imagined that the polyp could draw the juices out of this worm before it was in its stomach, but I soon became convinced that it could. The skin of the polyp being transparent, I could clearly see through it. The red fluid that it had drained from the worm upon which its lips were fastened was in the stomach of the polyp, or, if one prefers, in the funnel that the polyp formed. The volume of the fluid increased perceptibly as I watched.

Since then I have seen a number of polyps with prey in their arms that were too large to insert inside their stomachs; the polyps partially sucked out those animals by pressing their lips tightly against them.

The cause behind this suction may also lead to the detaching of the colored parts from the skin of the animals. I have tested this hypothesis using spiders [hydrachnids; see Book I, p. 48] of a very beautiful red color which are quite common in bodies of water. When I fed the polyps the skin of these spiders, the colored parts were separated out in the stomach of the polyps. After the coarsest parts of the skin had been detached and expelled, the red matter just described formed a red fluid in the stomach of the polyps that was very easy to discern through the polyps' skin.

The reddish substance from the water fleas produces a reddish fluid, and that of the flat white worms shown in Plate 7, Figure 7 gives a beautiful red. In short, the fluid remaining in the stomach of the polyp always shows the dominant color of the animal which the polyp has eaten.

This fluid consists of water present in the stomach of the polyp and of colored particles from the food. By squeezing the body of a polyp filled with fluid, one can see a rather large quantity of it come out.

I have noticed something which may contribute significantly to the digestion of the food, that is, the continual thrusting of the food back and forth from one end of the stomach to the other. In order to have an opportunity to observe this motion, it is necessary to select a polyp with a stomach that is not too full and which has ingested food that can be macerated. This tossing about of the food becomes perceptible once the parts have been divided into small fragments. To the contrary, no such motion is visible nor can it occur when the stomach of the polyp is very full and when the food which it has swallowed cannot be reduced into little pieces. In a space that is completely filled with tightly packed food, there is no free play allowing the pieces to be tumbled back and forth as occurs when the food is less crowded and more broken up. It is quite probable, however, that this motion always occurs in a like manner inside the polyp, but that it produces its effect on the food in a way that is not perceptible. It is quite natural to conclude that this tossing about of the food in the polyp's stomach is caused by a sort of peristaltic motion operating in several directions. One can readily imagine that in the polyps, as in other animals, this motion can contribute greatly to digestion.

The peristaltic motion distributes the nutritive substance throughout the stomach of the polyp. This does not seem particularly necessary in polyps of the second species, because their stomachs usually fill with food immediately from one end to the other. The situation differs in the polyps with long arms. As already shown, the body of this species narrows sharply at a point about two-thirds down the length of its body and remains narrow to the posterior tip. Consequently, in this section the stomach is greatly constricted. Most of the time the food swallowed by the long-armed polyp fills only the wider part of the stomach and stops at the point where the narrowing begins; the coarse fragments cannot enter. They must be separated out in order for the nutritive juice to be brought into the narrow portion of the stomach. This is accomplished by the peristaltic motion occurring in the polyps. To be certain, it suffices to give some polyps food, such as the flat worms with their intestines full of a red substance (Plate 7, Fig. 7) that furnishes rather brightly colored juices. Shortly after this red substance begins to come out of the victim's body and to spread through the wide part of the polyp's stomach, it is pushed next into the narrow section, reaching to its tip and collecting there in an appreciable quantity.

This tail section is not the only narrow passage into which food must be carried. There are a number of such narrow passages in the polyps of both species on which I have performed experiments. Such passages occur in the polyps' arms, which are hollow inside and which, like the body, form a kind of tube. The tubes in each arm connect with the tube formed by the body; that is, they open into the stomach. Food is driven by the peristaltic motion through this connecting cavity from the stomach into the arms.

I began to be persuaded of this fact on observing some polyps that had sucked the red substance out of the intestines of the flat worms just discussed. By using a magnifying glass to examine attentively the arms of these polyps near their base, I saw that the red substance had passed into the arms and that they contained an appreciable quantity of it. This red juice is composed, however, of particles too fine to be seen clearly enough for one to determine whether or not it passed through an opening linking the arms to the stomach and then into a hollow tube formed by the arms. The juice could have been filtered and could have diffused through the skin itself.

I performed an experiment which completely resolved my doubt. I had fed a polyp some pieces of skin of the little black flat slugs [*Polycelis;* see Book I, p. 48] (Plate 7, Fig. 9) that abound in ditches. In the stomach of the polyp, this material was soon reduced to a kind of pulp composed primarily of small black fragments. This pulp with these fragments was being tossed back and forth in the stomach. Aided by a good magnifying glass, I followed this movement attentively. Finally I perceived that a number of the small black fragments were passing single file into the arms of the polyp. I saw them distinctly both in the stomach and in the arms, and I saw them pass from the one into the other. I saw the fragments advance into the arm no further than the length of about five to seven millimeters when they were thrown back into the stomach and driven again toward the posterior extremity of the polyp. Afterwards they were pushed once more toward the anterior end and into the arms, and so on uninterruptedly.

I have repeated this experiment frequently. It seems to me to provide quite clear proof of this sort of peristaltic movement operating in the polyps in several directions. It likewise proves that it goes on in the arms as well as in the stomach. The motion is always rather slow, especially in winter.

The experiment I have just described seems to me to provide sufficient evidence that each arm of the polyp forms a tube which communicates with the stomach through an opening. I would not hazard a decision, however, as to whether or not this tube reaches to the end of the arm and whether or not the arm is open at its tip.

The preceding experiments were performed on polyps of the second and third species. They are carried out more easily with those of the third because their skin is more transparent than that of the other species.

I managed to see quite distinctly the openings which communicated between the stomach and the arms of the polyp. I have observed them many times, but the first time I saw them, I was not expecting to see them at all. I was performing an operation [p. 160] which it is not yet time to discuss.

From all that I have just said, it appears clear that once the nutrient substance has been extracted, it spreads throughout the tubes formed by the body and by each of the arms of the polyp. But how does the nutriment then pass into the skin which constitutes the walls of these tubes? How does it spread through its parts? And how does it contribute to the nutrition and growth of the polyps?

I will not promise to provide satisfactory answers to these questions. I am simply going to set forth some observations which perhaps will serve to shed a little light on the first two of them.

It is advisable that I begin with a digression on the source of the coloring of the polyp. The facts entailed will serve as a basis for my statements related to the preceding questions.

In the first *Memoir* (p. 26) I stated that the polyps are not always the same shade of color, that variations occur in an individual polyp. The color becomes sometimes deeper, sometimes lighter, or when it is lost entirely, leaves the polyp a slightly transparent white. All this is very easy to observe when the same polyps are examined successively for a certain period of time.

It is likewise very easy to perceive that these changes result from variations in the amount of food consumed by the polyps. The more they eat, the deeper their color becomes. When they are allowed to go without food, their color becomes lighter.

Polyps of the second and third species are ordinarily a reddish brown color, but the shade varies with the individual of these species, tending more or less to red or to brown.

After raising polyps for some time, I had reason to believe that this variation in color resulted not solely from the amount of food I was giving these animals, but also from the diversity in color of the food itself. Thus, I was led to conjecture that if the polyps would eat creatures more vividly colored than the millepedes and water fleas on which I had fed them until then, they would acquire a more pronounced coloration than that with which I was familiar. My first goal was to make them a beautiful red. For this I had recourse to those worms which have viscera full of a substance verging on crimson (Plate 7, Fig. 7). I fed some of these worms to several polyps of the third species that were nearly colorless, and within a day they were tinted red. I repeated the experiment on the same polyps to intensify the shade of their red coloring. I also performed it on other polyps of both species. I saw all the polyps take on the color of the substance in the viscera of the worms I had given them, the intensity varying in proportion to the amount of this substance which had passed into their stomachs.

After staining the polyps red, I aspired to color them black. For this purpose I fed them pieces of little black aquatic slugs (Plate 7, Fig. 9); in a short time the polyps became black. At various times it happened that the polyps did not digest the slugs very well; consequently they took on very little black color or none at all. To some I fed little black tadpoles of frogs; after digesting them, the bodies of the polyps took on a rather deep shade of black.

Those little aquatic spiders of an attractive red color were too conspicuous for me to overlook. Their color cost the lives of many that I fed to the polyps. I began by offering them intact to the polyps. Realizing that most polyps had difficulty in swallowing them, I fed the polyps only the skin that I removed from these spiders. In removing the skin, I learned at the same time that only the external surface was red. Polyps which partook of the skin of the spiders became a beautiful red, verging on the color of flame. At times I have found in the ditches polyps of this same color. It is probable that they had recently fed on red spiders.

I also undertook to stain polyps of the third species green. Finding no aquatic creature that could give them this color, I had recourse to a species of green aphids from rose bushes. I fed some to several polyps which, after swallowing and digesting them, took on a faint green tinge.

The rigid feet of these aphids make them difficult for the polyps to swallow. Most of the time I took the trouble to cut off their feet before feeding them to the polyps. I have seen a few polyps, however, swallow an aphid which had feet. The feet could not be accommodated inside the stomach; they forced the skin of the polyp to stretch to an extreme degree, and what is more, some feet perforated its skin and showed outside the body. A very peculiar effect is created by three or four of these feet emerging at various spots on the polyp's body.

Those perforated polyps did not digest the body of the aphids any the less; they even forced the feet back inside the stomach little by little. Thereafter no trace remained of the wounds made in the skin by the feet, and the polyps did not appear to suffer any ill effects at all.

After having tried to vary the color of the polyps by feeding them animals of different colors, I attempted to obtain such variation by yet another method. I conceived of nothing less than giving the polyps all the rich diversity of colors found in the flowers that grace a garden. For this scheme to succeed, these vegetable foods had to be appropriate for the polyps. I cut the petals of various flowers into narrow strips five to seven millimeters long and I dropped them on the arms of a number of polyps. Some did not seize them at all; others grasped them for a short while. As the polyps showed little haste in carrying the petals to their mouths, however, I soon understood that these flowers were not acceptable to them. Nevertheless, two polyps did swallow some: one a piece of petal of blue larkspur and the other, a piece of wallflower petal. As I was observing the polyps very attentively to learn whether or not they would digest them and separate out their colored parts from the rest, however, I saw the polyps expel them from their stomachs.

Thus I realized that polyps could not feed on flowers, and that I had to abandon hope of seeing this experiment succeed. I did not try to give other food of a vegetable nature to the polyps excepting some bread, on which they did not feed.

I tried placing polyps in water in which yellow marigolds and blue larkspur had been steeped. I was testing whether or not they could survive in this fluid, and especially whether or not the colored substances in the liquid, passing into their stomachs, would act at all to nourish and tint them as does the colored matter they extract from animal food. The polyps, however, were unable to live in these infusions.

From the experiments reported above, it thus would appear that the color of polyps is derived from that of the nutritive juice they extract from the animals they eat. One might suppose, perhaps, that the polyps only give the appearance of being colored by their food, just as a bottle of transparent glass seems to be the color of the liquid it contains. To be convinced to the contrary, one need only examine with care a polyp that has completely voided all its wastes and that has elongated once again. One can see with the naked eye that the stomach of the polyp has shrunk greatly, that it no longer contains any fluid and that the color is in the skin itself. It is still easier to become convinced if one then cuts the polyp open.

Wishing to learn how long the polyps would retain their color, I experimented with some red and with some black ones, all of the same species. After they had digested the food that had colored them, I allowed them to go unfed. At the end of fifteen days, I found that these polyps still were a quite recognizable shade of black or red, although much paler than they had been immediately after eating. Gradually the color weakened so much that at the end of three weeks or a month it was no longer visible; the polyp was nearly white.

But how does this colored substance spread into the skin of the polyp? Does it enter vessels and is it dispensed via these vessels throughout the entire body? Let us recall here that I have not been able to discover a single vessel in the skin of the polyp, as I mentioned in the first *Memoir* (p. 31). I have seen only those granules which I have discussed a great deal and the viscous material encompassing them. It is possible that this material does contain a number of vessels. But even supposing this to be the case, I venture to state nevertheless that the colored substance, upon being drawn out of the food, does not immediately pass into any such vessels. For this viscous material always remains white and transparent, even in the polyps having the deepest coloring.

To the contrary, the granules that abound in the polyp's skin are colored; it is on their color that the color of the polyp depends, and they, in turn, take their color from the nutrient substance that has been extracted from the food ingested by the polyp. The granules become red or black, for example, when the polyps have been fed a red or black substance. They show varying intensities of these different colors in proportion to the quantity and the depth of color of the nutrient juice. In the end, the granules lose their color gradually unless one maintains it by giving the polyps food of the same color from time to time. What then may be concluded from all this and from all that was said previously on this subject (*Memoir* I, p. 32 and following)? That the nutritive substance passes immediately from the stomach of the polyp into the granules which fill the skin.

But how does the substance color the granules? It is quite plain that the granules are all either so many small *glands* or *vesicles* into which the substance is brought, perhaps by a kind of suction. These vesicles appear colored because they are full of the colored substance they have taken in. Based on this assumption, it is easy to understand why these granules do not appear colored when they are separated out, and why they appear more and more deeply colored in proportion to the quantities of them that are grouped together.

As I have said in the preceding *Memoir* (p. 30), the walls of the stomach are completely lined with granules. These granules are the first to be filled with the nutrient substance that is in the stomach. Accordingly, after the polyps have eaten, the inner surface of the stomach is always well colored. The layers of granules above this surface also absorb the nutrient substance, the amount varying with the volume of food taken by the polyp.

When the polyp has eaten little, and especially when it has been without food, one readily notices that only a few layers of granules are colored, namely, those closest to the walls of the stomach. The other layers of granules are white, forming the transparent envelope filled with colorless granules that I have discussed (pp. 26-27). When a polyp is well fed, the transparent envelope is thinner and more layers of granules are colored. Undoubtedly, as the granules, that is the little glands, nearest the stomach become filled, the nutrient substance passes to the others. When one recalls that the structure of the arms is the same as that of the body and that the nutrient substance penetrates into their interior (see p. 34 and following and p. 76 and following), it is easy to understand that the arms draw in the substance in the same way as does the body.

Count Marsigli in his *Histoire de la Mer* informs us that the skin of the coral and of other so-called marine plants is completely filled with little granules which he compares to grains of salt. The preface to the sixth volume of the *Mémoires pour servir à l'Histoire des Insectes* contains the fine observations that Mr. Réaumur made jointly with Mr. Bernard de Jussieu and Mr. Guettard. Now these observations reliably inform us that those organisms that were taken for plants are masses of polyps in polyparies. The granules found in those polyps should be studied, for if they resemble those found in the polyps whose history I am presenting, but are larger, we may be better able to learn what these granules are.

At this juncture one might ask, what becomes of the nutrient substance after it has passed into these granules? How is it distributed to nourish the other parts of the polyp? I find myself completely unable to answer this question.

Polyps grow rapidly when they eat abundantly and often, that is to say, in the summer. Growth is proportional to the amount of food they consume. I will not tarry at this time to point out the various stages of their growth because the subject will arise again when I discuss the manner by which the polyps multiply.

I have previously stated that polyps are able to go without food for a very long time. Here I must add that they grow smaller as they fast. I observed some which had been an inch long when elongated, diminish in size until they became the smallest polyps I have ever seen; this resulted from a fast of more than three months duration. They grow smaller more quickly in summer than in winter.

In order to ascertain the life span of the polyps so far as I was able, I undertook to maintain some with continuity. I have some born in June and July of 1741 that are still living at the present time (January, 1744). Thus, these animals can live for more than two years. The ones I just mentioned have undergone many variations in size during their lifetimes. I have seen them now bigger, now smaller a number of times, depending on the amount of food I gave them.

Only by careful tending can polyps be preserved; even then many die of various maladies. I have already described one of them previously (p. 32) for which I could not find a cause.

Polyps are prone to be troubled by a kind of small lice [*Kerona;* see Book I, p. 48] very common in waters. These lice appear to have a flat underbody and a rounded upper portion. Their shape is almost oval. Ordinarily they are white, although with the aid of a magnifying glass I have noticed some brown coloring on the bodies of a number of them. They move about rapidly on the bodies of the polyps, and they can also slip off the polyps and set about swimming. They seem to me to collect chiefly near the head of the polyp (Plate 7, Fig. 10, *a*), although large numbers are frequently seen extending over the entire body *a b* and the arms *c, c, c.* A magnifying glass must be used to see them clearly. Figure 10 represents exactly how a polyp infested with these little animals appears through a strong magnifying glass. If polyps are not freed of these lice, they can become covered with them in a few days, and it is soon apparent that they are ailing because of them. Often polyps die and are eaten by these little animals. Gradually their arms diminish, and then their bodies, until nothing more remains of them. At other times they are consumed only partially, the head alone being lost (Plate 7, Fig. 11).

I have often put hundreds of polyps infested with lice into jars and purposely abandoned them to their sorry fate. The lice multiplied prodigiously in these jars and caused many polyps to perish. Occasionally all the lice disappeared after a certain time. I should point out that I never changed the water in these jars. The polyps that survived until the lice were gone continued, for the most part, to remain alive. Those polyps which had their anterior portion completely eaten away (Fig. 11) grew a new head and new arms. I nurtured some of these as well as some others which had not been quite so mutilated, and they became very handsome polyps.

Since polyps rarely exist for long without contracting lice, I was compelled to take a number of precautions against the infestation.

Even in the ditches where they dwell polyps contract lice. Many times I have found them heavily encrusted with lice, and I was able to verify that the polyps die of them in their natural habitat as well as in the powder jars.

Polyps that are covered with a great number of lice seem to lose their color more rapidly than the others. In addition, I have often noticed that although the infested polyps were fed as much as the others, they did not take on as deep a color.

Experience taught me that it was good to change the water of the polyps frequently. It is especially important to put the sick ones into fresh water. Ordinarily I have followed the practice of changing the water after each meal the polyps eat to prevent the water from becoming contaminated by their excrements. I have seen polyps appear to be putrefying when left in water heavily contaminated by the cadavers of water fleas and other wastes.

When polyps are especially valuable to me because of experiments that I have performed or that I am in the process of performing on them, I take still further precautions. I try to free them of their lice entirely by passing the tip of a brush over their whole body many times. When I cannot conveniently brush the polyps, I shake them in the water, a procedure that also helps dislodge the lice.

Few creatures lack enemies. Waters teem with voracious animals destroying one another. One which has devoured many is devoured, in its turn, often by animals much smaller than itself. The polyp is an example. It is perilous to worms which greatly exceed it in size, devouring them alive; yet tiny creatures, not even clearly visible without a magnifying glass, attack the polyp and cause its death.

I searched the waters for larger animals that would eat polyps. First I offered polyps to some fishes. I threw a polyp into a glass vessel containing a perch of a small-sized variety which I had been feeding on worms for some time. The fish approached the polyp at once and gulped it in. Instead of swallowing it, however, the perch immediately spit it out as if it found the polyp extremely repugnant. Once again the perch drew near, gulped the polyp in anew and rejected it with the same speed as before. Then the episode recurred five or six times in succession. I repeated the experiment with the same perch many times. As often as I offered it polyps, the perch invariably undertook to swallow them. Except for two times when it did succeed, however, it always rejected them the very instant after taking them into its mouth.

I offered it a polyp which held in its arms a worm greatly relished by the perch. I conjectured that in order to swallow the worm, the fish perhaps would be compelled to swallow the polyp. At first, it did indeed gulp in both the worm and the polyp, but instantly spit them out. Subsequently the perch seized and rejected them repeatedly until the two were separated; thereupon it swallowed the worm and left the polyp. I have often repeated this experiment in the presence of various people who were highly amused by the performance.

I carried out the same experiment using gudgeons and obtained the same results. Instead of spewing the polyps out of their mouths, they ejected them through their gills a number of times. Of the three gudgeons to which I frequently offered polyps, only a single fish once swallowed one. Ordinarily the fishes reject the polyps without doing them the least harm, but I did see one polyp badly torn and barely recognizable after the fish had attempted to swallow it a number of times. This accident which would have been fatal for many other animals was of minor consequence for this polyp; it soon healed.

I have not been able to find any species of aquatic beetles that would feed on polyps although it is common knowledge that some beetles are extremely voracious. I tried particularly to get those of a rather common species [a whirligig; see Book I, p. 48] to eat the polyps. These were the small, nearly oval beetles whose wing casings have the color and luster of burnished steel. They are often seen swimming in a throng across the surface of the water at an astonishing speed. I collected a number of them in a large glass vessel where I fed them various terrestrial and aquatic creatures. When they are hungry, they seize with haste those prey I have placed on the surface of the water, fighting over them amongst themselves quite violently. After keeping them without food for some time, I offered them a polyp. It was seized at once by two of these beetles, but a moment later they released it. I have seen this happen many times. No matter how hungry they were, they invariably rejected the polyps.

I gave a large worm to a small polyp. When it had swallowed the entire worm and was extremely swollen, I offered the polyp to my beetles. Several beetles seized it and then ate the polyp together with the worm inside it. This is the only way I have been able to make the beetles eat polyps. It seems to me, however, that this case does not prove at all that the polyps are a natural food for them.

I pointed out my techniques for finding polyps earlier in the first *Memoir*. There I said that they are to be found indiscriminately upon all manner of things in the water. These objects should be removed from the pond and placed in glass containers filled with water. If the vessel is left undisturbed for a moment, the polyps will have time to elongate and will soon be noticeable, particularly to someone who already has had the experience of seeing them.

The ditch where I originally found polyps of the first species, and subsequently specimens of the other two species that I discovered, is located at the foot of some dunes. It is about twenty feet wide. At one end it closes into a cove. At the other it connects through a narrow channel to a stream which runs lengthwise through Sorgvliet. The water in this ditch is continually freshened by water filtering through the sand of the dunes and flowing from this ditch into the aforementioned stream. Without being stagnant in the least, the water in the ditch is clear to the bottom and teems with creatures. As I shall describe later, I saw an extraordinary number of polyps there.

I searched for them in the other waters of Sorgvliet which are connected to the ditch, such as a fishpond traversed by the stream and the stream itself. I came upon polyps, especially those of the third species, in various locations. I surveyed the stream to its borders above Sorgvliet, and I fished for polyps even in places where the water was only a few inches deep. Finally, I found polyps in large ditches closer to The Hague than to Sorgvliet, and I have every reason to believe that they inhabit most of the canals and ditches of Holland.

The polyps seen by Leeuwenhoek were taken from the canal that goes from Delft to Delftshaven. Mr. Allamand, who was kind enough to repeat most of my experiments and whose testimony I can cite as an excellent proof of the facts I discovered, found great quantities of polyps in the Province of Friesland and in the vicinity of Leiden. In July, 1740, prodigious numbers of them were collected from the ditches around a country house located in the Province of Overissel. Mr. de Réaumur found three different species around Paris. These animals have been seen in England since 1703, as we are informed by the anonymous letter in the *Philosophical Transactions* from that year. Since April of the year 1743, polyps have been caught at Hackney, near London, and subsequently in a number of areas of England and Scotland.

There is reason to think, therefore, that most naturalists who may wish to verify and to improve on my experiments will easily find polyps in the various countries of Europe. They should be sought especially in the corners of ditches, ponds, and pools in places where the wind blows and accumulates floating plants. Though one may have searched without success in such places, he should return, for he may find a great many polyps there a week or two after he was unable to uncover a single one. This I know from experience.

Polyps are much scarcer in the waters during winter than in the other seasons of the year, and they are more difficult to find then. In this season, the aquatic plants to which the polyps ordinarily affix themselves either no longer float on the surface of the water or rise from the bottom only in small numbers. Most are annual plants which decay as winter approaches, their remains sinking to the bottom of the water. The polyps are also at the bottom, but in a kind of inactive state. Before one can find a single polyp, it often requires the patience to draw many objects out of the water and to examine them well. I have found polyps in every month of winter. But when the fine weather approaches beginning in early April, the duckweed climbs to the surface of the water and multiplies there, young horsetails sprout and rise up in the water and many other floating plants begin to grow. The polyps, revived by the warmth, climb upon these plants and move over them while seeking their prey. Then the polyps can be drawn from the water together with the plants.

As the warmth increases, one is likely to find polyps in larger quantities, if not in one place then in another.

I stated that in winter especially the polyps stay at the bottom of the water.

I learned this during that season both from the research which I carried out in the ditches and from experiments which I performed in my study. I kept a number of polyps in large vessels, the bottoms of which were covered with earth. They remained on this earth at the bottom of the water while the weather was cold, but as soon as they felt the warmth, they climbed along the walls of the vessel, and from the walls onto the aquatic plants which I had placed there and to the surface of the water. It would be dangerous for the polyps to remain near the surface in freezing temperatures as they would risk perishing in the ice. I tried to freeze them. Some remained in the ice for more than twenty-four hours and did not die, but a larger number perished. Consequently, it is natural to think that such a sojourn in the ice, especially when it is of rather long duration, could only be destructive to the polyps. Therefore, it is extremely important to take precautions to preserve polyps from freezing when they are being used in experiments.

During one hard freeze, I saw polyps through the ice in the ditch where I usually fish for them. To observe them, I stretched out on the ice in places where the sun shone and illuminated the water down to the bottom. Polyps both of the second and third species were fastened onto the plants and leaves at the bottom of the ditch. I made my most careful study of these polyps on January 11 of this year, 1744, when the thermometer of Prins, exposed to the north, read 16 degrees in the morning, and 26 at noon. Observing the polyps between noon and one o'clock, I saw that they were not extremely contracted, and I noticed one that had eaten. Although the water does not teem with creatures in winter as it does in summer, some are nonetheless visible even in the coldest weather. On that particular day, for example, I noticed three different species of water fleas swimming about which could serve as food for the polyps. I was curious to know how cold it was around the polyps. After making a hole in the ice, I lowered a thermometer suspended on a string to the bottom of the pond and left it for some time alongside the polyps. When I drew it out, it read 42 degrees. Cold does not penetrate as quickly to the bottom in a ditch as it does in a powder jar. Thus, in their natural habitat the polyps are not readily exposed to the cold to which I subjected them in my jars, a cold which numbed them entirely and completely took away their appetite. I have always kept my polyps in well water because it was the easiest for me to obtain.

I have also kept them in rain water. Water from springs and rivers ought to be quite suitable for them. It is likely enough that certain waters are not fit for the polyps. Often polyps placed in a certain type of water all die in a short time, although this does not prove that the water is unsuitable for these animals. By taking some precautions, one may subsequently demonstrate that they can live in it.

EXPLANATION OF

THE FIGURES OF THE SECOND MEMOIR

PLATE 6

FIGURE 1 represents a barbed millepede [*Stylaria;* see Book I, p. 48]; *d* shows the barb at the anterior end of this animal.

Figure 2 portrays a polyp *a b* of the second species. A portion *a m* of a millipede extends from its mouth. The other portion is in the stomach of the polyp.

Figure 3 shows a piece of string to which a polyp *a b* is attached at point *b*. The string should be placed across the top of a large powder jar so that *h* and *k* rest on the edges of the glass while section *b* dips into the water. Then there is room for the arms of the polyp to hang downward; when they are extended one is able to see the polyp's maneuvers in seizing its prey and carrying it to its mouth. One millepede *m c n* is grasped by the end of the arm *a c* and is dragging the arm along with it as it swims. Another millepede marked *m i n* is already very much entangled in the arm *a o i*, and the polyp is in the process of carrying it to its mouth. For this purpose, the polyp not only contracts its arm but also twists it into a corkscrew from *o* to *i*. A water flea *p* has been grasped by the tip of the arm *a p* and is dragging this arm along as it struggles.

Figure 4 shows a polyp that has seized a large worm and is preparing to swallow it while the worm is folded in half. The lips of the polyp are already stretched wide open. The worm is thoroughly entangled in the arms of the polyp.

In Figure 5 a long-armed polyp is depicted having just swallowed a worm which is quite recognizable through the skin of the polyp. The end portions of the two arms that had clasped the worm have been swallowed along with the prey, and the loops they form are seen at *i* and *e*.

In Figure 6 a polyp is portrayed swallowing the middle of a doubled-over worm. The two ends of the worm are still hanging outside the mouth of the polyp.

Figure 7 shows the polyp's mouth *a* open and a portion of a water flea *p* already inside it.

The polyp of the second species portrayed in Figure 8 is entirely filled with water fleas.

Figure 9 shows a long-armed polyp. The portion *a d* of its body is filled with water fleas whereas the tail *d b* is empty and very narrow. There is a neck at *a*.

The polyp depicted in Figure 10 is similar to the one shown in Figure 9; since it is not so completely filled with water fleas near the mouth as is the other, the neck *c* is still more prominent.

Figure 11 illustrates a water flea with branched horns [*Daphnia;* see Book I, p. 48] enlarged under the microscope; *b, b* are its branches or antennae, *c* is its anterior end, and *e* is its tail.

PLATE 7

FIGURE 1 depicts a long-armed polyp that has seized a great many water fleas and has gathered them all into a clump *a* near its mouth in order to swallow them one after the other until its stomach is filled.

Figure 2 shows the worms frequently found in large masses at the bottom of ponds. The portions of their bodies held above the earth undulate continually.

In Figure 3 a polyp of the third species is shown hanging attached to a twig by its posterior end *b*. It has swallowed a roach fry. The skin of the polyp is stretched so thin that it can scarcely be discerned. Clearly distinguishable are only the fish and the polyp's arms *a c*, *a c*, *a c*, which could be mistaken for barbules emerging from the head of the roach fry.

Figure 4 depicts two polyps, each of which has swallowed the opposite end of the same worm so that their heads have met at *a*. Sometimes one of the polyps will open its mouth wider and partially swallow the other. This occurrence is represented in Figure 5.

Figure 6 shows a polyp discharging excrement *e* composed of the remains of a worm that has not been completely digested.

Figure 7 pictures a flat white worm [*Glossiphonia;* see Book I, p. 48] that has viscera filled with material of a beautiful red color.

Figure 8 represents the red larva *c d* of a midge [*Chironomus;* see Book I, p. 48]. The polyp has fastened its lips *e i* to the victim and is sucking the nutrients from it in this position without swallowing it.

A black aquatic slug [*Polycelis;* see Book I, p. 48] is shown in Figure 9.

In Figure 10 a polyp fastened to a twig by its posterior end *b* is portrayed enlarged as seen through a microscope. Many tiny white lice cling to the polyp's body *a b* and to its arms *a c*, *a c, a c,* but especially to its anterior end *a*. They come and go sucking on the polyp, possibly causing its death.

The lice [*Kerona;* see Book I, p. 48] have already eaten the arms and head of the polyp depicted in Figure 11.

Pl. 6. Mem. 2.

Fig. 1.

Fig. 2.

Fig. 3.

Fig. 4.

Fig. 5.

Fig. 11.

Fig. 10.

Fig. 9.

Fig. 8.

Fig. 7.

Fig. 6.

P. Lyonet delin. et sculp. 1743.

MÉMOIRES
POUR L'HISTOIRE
DES POLYPES.

****

TROISIÉME MÉMOIRE.
De la Génération des Polypes.

PRÈS m'être affuré, que les Polypes d'eau douce, dont il eft queftion dans ces Mémoires, pouvoient fe multiplier par la fection, je fus extrêmement curieux de connoitre la maniére dont ils fe multiplient naturelle-ment. Je doutois encore, dans ce tems-là, fi les Po-lypes étoient des Plantes, ou des Animaux, & je
me

Pl. 7. Mem. 2.

Fig. 1.

Fig. 2.

Fig. 3.

a

b

a

c c c

Fig. 6.

Fig. 5.

Fig. 4.

a

e

Fig. 9.

Fig. 7.

Fig. 10.

b

Fig. 11.

Fig. 8.

b

d

c

a

c

c

c

P. Lyonet delin. et sculp. 1743.

MEMOIRS

CONCERNING THE NATURAL HISTORY

OF THE POLYPS.

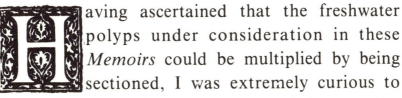

THIRD MEMOIR.

On the reproduction of polyps.

aving ascertained that the freshwater polyps under consideration in these *Memoirs* could be multiplied by being sectioned, I was extremely curious to know the manner in which they multiply naturally. At that time I was still undecided whether the polyps were plants or animals; and, as I have said, I even thought it likely that their natural mode of multiplication would provide me with a sufficiently distinctive characteristic to free me from doubt.

The letters of Leeuwenhoek and of the anonymous Englishman, printed in the *Philosophical Transactions* for the year 1703, indicate that the first and apparently the only polyps that these two naturalists observed were in the process of multiplying when they were found. However novel the extraordinary phenomenon of the polyps' multiplication may have been for them, they did not misinterpret it at all. Struck by this marvel, they gave all their attention to the progress of the young polyps until they were completely formed and had separated from their mother.

I was unaware of the discoveries made by the two naturalists when chance introduced me to the polyps, and at first I did not recognize their offspring for what they were, although several among the first polyps I discovered were producing young. I scarcely paused to consider them, as I had not yet made any plans at all to observe the polyps systematically. When I did set about examining them carefully, there were no longer any polyps among those in my glass vessel that were in the process of multiplying.

It was in the month of December, 1740 that I began to explore how the polyps multiply naturally. I did not perceive anything in December, January, or part of February of 1741, however, to encourage the hope that I would succeed in satisfying my curiosity on the subject. I could hardly count on success in winter, but impatiently awaited the season favorable for plants to proliferate and animals to multiply. I found what I was seeking even sooner than I had hoped. On February 25, I was examining the vessel in which I put all those polyps not being used in any particular experiment. Looking through a magnifying glass I observed a green polyp which was fastened to the side of the glass and which had a small dark green excrescence on its body. However minute the excrescence, it drew all my attention because even though I had already observed quite a large number of polyps, I had never seen anything like it on any of them. From that moment the polyp bearing it was very precious to me, and from fear of confusing it with a number of others that were in the same container, I took it out and placed it by itself in another jar. By good fortune, it attached itself to the side of the jar in such a manner that, aided by a strong magnifying glass, I could easily observe the excrescence on its body. On the very day I discovered it, after scrutinizing it again and again, I found some similarity between its structure and that of the polyp from which it protruded. Certainly, I could not base any confident decision on such a vague resemblance, but I began to suspect that the excrescence would become a polyp. In thinking that I was watching a young polyp emerge out of another, I was especially influenced by my recollection of having seen polyps fastened on one another the preceding summer. I ceased my observations of February 25th at ten-thirty in the evening; I was impatient to resume watching this polyp which had occupied me a good part of the day.

The following morning I had the pleasure of seeing that the excrescence had increased in length. It was about a half a millimeter long, almost cylindrical, and was situated perpendicularly on the polyp. It grew noticeably larger on the 26th and the 27th until on the 28th it was at least a millimeter long. When I made the polyp contract, the excrescence also contracted. In short, all the evidence was tending to persuade me that this excrescence was a young polyp. To be certain of it, I needed only to see those fine threads, which I have called the arms and legs of the polyp, emerging from its top end. I kept looking for them, and on this same day, the 28th of February, at ten o'clock in the evening, I saw four beginning to sprout. The sight of them gave me appreciable pleasure; it greatly excited my curiosity to witness the end of a phenomenon so new to me. During the following days the four arms of the young polyp lengthened and a fifth appeared. The polyp itself grew markedly, stretching to a length of at least seven millimeters while it was still attached to its mother. It separated from her on the 18th of March between ten and eleven o'clock in the morning.

Before this young polyp had detached from its mother, I saw others also emerging from some of the group of polyps remaining in my large glass vessel.

In April, 1741 I found the second species of polyp discussed in the preceding *Memoirs*. The first individual of this species which I saw had five little ones emerging from its body at the same time. The long-armed polyps also were producing young when I came upon them. The systematic observations which I made on all the species proved to me that they all multiply in the same manner.

Because the polyps of the second and third species are much larger than the green polyps, it is easier to observe the progress of their offspring; hence I often repeated on them the kind of observations that I first made on the green species.

When a young polyp begins to grow, at first only a small and usually pointed excrescence is visible (Plate 8, Fig. 1, *e*). The excrescence is shaped rather like a short cone with a wide base. The color of this excrescence, or bud, is ordinarily deeper than the color of the mother's body. Little by little this bud springs forth further, and as it grows longer the cone changes its shape, becoming smaller at the base in proportion to the increase in height. Often the cone is misshapen, its tip either rounded or truncated instead of pointed. After the young polyp grows somewhat, it finally loses its conical shape and becomes almost cylindrical. At this time or thereabouts the arms begin to sprout at its anterior end (Plate 8, Fig. 1, *c i*). This young polyp does not maintain its cylindrical shape for long. Its posterior end, by which it holds to the mother, gradually narrows, constricts, and finally appears to touch the mother only at one point (Fig. 2, *b*). The body of the young polyp, widest by far at its posterior end during the beginning stages of its development, is nowhere else as slender once it is fully formed.

During the period when a young polyp has not yet developed very much, and while its posterior end is constricted only a little or not at all, it always remains in the same location relative to the body of its mother. It retains the same position it had on emerging from its mother, which is most often perpendicular to her, but sometimes at a more or less oblique angle. Once the constriction, as shown in Figure 2, *b*, has become appreciable, and the young polyp is fastened less to its mother, it no longer stays in a fixed position on the mother, but bends from her body at various angles. If at this time the young one were not supporting itself by its own power, it would hang downward. It is evident that it is supporting itself, however, and that it changes the direction of its body by executing all the maneuvers made by a polyp which is fastened against the walls of a vessel.

The polyp is ready to separate from its mother when it is no longer attached to her except at a single point (Fig. 2, *a b*). It is easy enough to imagine how this separation takes place, but I wanted to see it. By dint of keeping watch for the right moment, I have seen it happen a number of times. I have seen young polyps of all the three species that I know detach from their mothers, and they all separate in the same manner. Both the mother and the young one fasten themselves to the glass or to other objects on which they happen to be, using their arms, or even their heads. To separate, they need merely contract and not necessarily a great deal. Sometimes it is the mother that does the straining, sometimes the young one, and often both of them together. Recall that polyps are capable of many gradations of contraction, extension, and inflection, and that they can attach themselves to objects while they are in most of the quite varied postures which they are capable of assuming. Thus, it is easy to understand that when mothers and young prepare to separate, they do not always do so by assuming precisely the same position. To the contrary, the position varies greatly, but basically the process always corresponds to the general patterns that I have presented. I will be content to describe one particular case.

With its young polyp (Plate 8, Fig. 3, *c d*) ready to separate, a mother polyp (Fig. 3, *a b*) twisted its body into a circular arc *a d b*, fastening its two extremities to the sides of the powder jar. The young polyp held on to the mother at the high point of the arc *d,* while its head *c* fastened against the glass. The mother had only to contract its body a little to become straightened (Fig. 4, *a b*) into the chord of the circular arc it had constituted just beforehand (Fig. 3, *a b*). Its extremities remained fastened to the glass. The young polyp, clinging to the glass with its arms, did not move at all with its mother as she contracted, but remained in place. Thus its posterior tip *d* detached from the body of the mother polyp *a b*. They separated and, right from the start, turned out to be some distance away from each other (Plate 8, Fig. 4).

I have seen this type of separation occur even though the mother polyp was attached only by its posterior end. Separations take place even when both the ends adhere to nothing more than the surface of the water; the polyps find sufficient support there. I have often hastened the separation by striking the young polyp lightly with the tip of a brush or by shaking the mother in the water. On one occasion such shaking detached three young from a mother polyp bearing a number of them. At another time, I happened to engage the arms of a young polyp at the end of a brush and to hold it suspended in the water. I proceeded to shake it, and with it, the mother hanging down below it. They then separated from each other, the mother falling to the bottom of the jar, and the young one remaining attached to the end of the brush.

As soon as a young polyp separates from its mother, it can walk and perform all the movements of which these animals are capable.

The arms begin to grow, as I have already mentioned, at approximately the time that the young polyps assume a cylindrical shape (Plate 8, Fig. 1, *c i*). I say approximately because these arms, nevertheless, can emerge earlier or later, and because the changes of shape noticeable in young polyps during their growth are not always the same, nor do the arms always emerge in exactly the same order.

At first only four or five arms appear. When these arms have developed somewhat, others emerge in the spaces between them.

The polyps of a particular species do not all have the same number of arms when they separate from their mothers. After separating, they develop additional arms, their number varying with the individual polyps.

In the first *Memoir* (p. 13), I remarked that in none of the three species known to me do all the individuals have an equal number of arms, and I noted as precisely as I could the variations found among these animals in this respect. It was on polyps of the second species that I systematically observed the increase in the number of arms, because they are the ones I have maintained the longest. In some I have seen the number of arms increase more than a year after the birth of these polyps, the total at times even gradually reaching eighteen (Plate 10, Fig. 3) or twenty. I never could find any polyps in the ditches that had such a large number of arms; I have noted them only among those polyps that I have raised. I have also observed that the number of their arms sometimes decreases.

Even before they have separated from their mother, young polyps can use their arms to seize prey. To verify this fact quite easily, one need only drop a worm or some other small animal on young polyps still attached to their mothers. These young polyps will seize the prey even though they have as yet only a few arms and even though these arms are not yet full grown. This experiment also shows that the mouth of a young polyp is formed before it separates. The young polyp opens its mouth to insert the worm into its stomach. In short and generally speaking, it does everything that I described in the preceding *Memoir* as constituting the means by which a polyp seizes prey, carries it to the mouth, and swallows it.

The length of time it takes for young polyps to develop varies greatly and depends on the temperature and on the amount of food they consume. I have seen young polyps fully formed at the end of 24 hours and others still immature after two weeks. The former grew during the height of summer, the latter during a time when the water around them registered below 48 degrees on the Prins thermometer, a temperature lower than moderate.

The same conditions that cause the young polyps to grow more or less rapidly also influence the length of time they remain attached to the mother. The faster they form, the sooner they are ready to separate. Accordingly, in summer, one very often sees young polyps detach from their mothers two days after starting to grow, whereas in winter five or six weeks sometimes elapse before they separate.

The size of young polyps when they separate from their mothers varies depending on the quantity of food they have received during the union and on diverse conditions which it would be useless to specify.

A young polyp does not necessarily always leave the mother as soon as it is well formed and its posterior end, by which it is united to the mother, has thoroughly constricted. If the mother has settled in a location where she and her young are always able to seize as much prey as they can eat, the separation, even in the warmest weather, is delayed for several days after the young have reached the point where they could separate. When food is scarce, however, the young detach sooner. Apparently, hunger drives them elsewhere to seek the means to satisfy it.

I shall shortly report my observations on the degree of warmth required for the development of young polyps and on the duration of their union with their mother. Hence, I shall not enlarge on these matters any further for the present.

It could be said, perhaps, that the young polyps do not really emerge from the body of the parent, but that eggs or little polyps have simply been deposited and fastened on the skin where they remain attached while they grow. This supposition would occur quite naturally to those who have not examined the phenomenon themselves, and it could arise also in the minds of those examining it for the first time. After carefully scrutinizing a young polyp through a magnifying glass, however, and especially after following its development, the observer can scarcely doubt any longer that the mother puts forth the young out of her own body as the trunk of a tree puts forth a branch. It can be seen distinctly that the excrescence which is the beginning of a new polyp is nothing other than a continuation of the mother's skin which has swollen and risen in that spot. The observer can even discern that the excrescence already forms within it a small tubular cavity which communicates with the tube shaped by the mother's skin, or in other words, with her stomach.

My initial observations taught me that much, but I was far too interested in ascertaining the facts more positively to stop there. Thus, I spared no pains to perform more decisive experiments regarding the nature of the union between the young polyps and their mothers.

It was above all a question of knowing whether or not the stomachs of the offspring communicate with those of the mothers. My earlier comments on the structure of the polyps (p. 28) should be recalled here. Their bodies are hollow from one end to the other, forming a kind of tube or gut which I have called the polyp's stomach. In order to ascertain the nature of the narrow connection which appeared to exist between the mother and the young, I therefore proposed to investigate whether or not the tubular cavities formed by their bodies had a connecting passage between them. Were the stomachs of the mother and the young joined in a fashion similar to that of certain vessels of the human body which open into each other?

My first experiment with this end in mind consisted of attempting to open a polyp in such a way that I could distinctly see the connecting passage, presuming one existed, between the stomach of the mother and that of the young polyp. I chose a rather large polyp of the second species from which a young one was emerging. At the posterior end of the new polyp, the part by which it was attached to the mother had not yet begun to constrict. Given these factors, if there were a communicating passage, it would be all the more noticeable. I placed the polyp into a little water in the hollow of my hand and stationed myself so that I could, when I wished, obtain more light on the specimen by exposing it to the sun. Next, I cut off about half of the young polyp's body with a scissors, opening its stomach to view. At the upper end of the portion still attached to the mother there was now an opening through which I could look by using a magnifying glass. It seemed to me that the stomach of the young polyp did indeed connect with that of the mother. I could have been deceiving myself, however. For greater certainty, I needed more light to see the area where the connection ought to be. Therefore I cut through the mother's body on both sides of the spot where the young polyp had protruded, leaving only a very short cylindrical portion open at both ends. These two cut openings served to bring more light into the remaining portion of the mother's body and consequently to better illuminate the spot where the opening communicating between the mother and the young one ought to be. I looked anew through the upper open end of the remaining portion of the young polyp and I saw noticeably more light at the spot where the young one was attached to the mother; it even seemed to me that I saw into the stomach of the mother. It was still possible, however, that at the place where the two polyps joined there could be a skin which did not impede the passage of light but which nevertheless separated the two stomachs. This possibility remained to be investigated.

With this aim in mind, I cut the remaining cylindrical portion of the mother lengthwise and removed the half opposite that from which the young one projected. This operation exposed the area on the lining of the mother's stomach where the hole connecting it to the stomach of the young one ought to be. Then I saw it quite distinctly (Plate 8, Fig. 5, *t*), and looking through it with a magnifying glass, I observed the opening (Fig. 5, *o*) at the end of the remaining portion of the young polyp. Next, reversing the position of the two sections of polyps prepared in this way, I looked through the opening on the young polyp (Fig. 6, *e*) and very clearly saw light through the connecting passage (Fig. 6, *i*). In order to have no remaining qualms on the matter, I placed these portions of polyps into a small shallow glass and, through a magnifying glass, observed them again very attentively. Positioning them in the glass as I had already done in my hand, I saw the connecting opening that I was seeking so clearly that I no longer had the least reason to question its existence.

I was not satisfied with doing this experiment once; I tried it on seven occasions and succeeded on five.

It is scarcely possible to feed polyps when they are producing young without having occasion to notice a fact which demonstrates the existence of this connecting opening between the stomachs of the mother and her young. After the mother has eaten (Plate 8, Fig. 7, *a b*), one sees the bodies of the young polyps that are attached to her swell, filling up with food as though they had taken it themselves through their own mouths (Fig. 7, c, d, e, i, o). Even a cursory look affords one every reason to believe that the food has passed from the stomach of the mother into theirs through the connecting opening between them. The proof becomes still more complete, however, if one can actually watch food passing from the stomach of the mother into the stomachs of her young. With a little effort, one will easily find a chance to see this take place.

I was raising an isolated long-armed polyp in one of my powder jars; from it a young one was emerging at a point about two millimeters away from its head. This mother polyp was holding fast to the wall of the jar, her head pointing downwards. At a quick glance, there appeared to be a very large connecting opening between this polyp and its offspring. Together the two polyps resembled a tube which bifurcates quite near one of its extremities. To ascertain the nature of this structure more precisely, I gave a barbed millepede [*Stylaria;* see Book I, p. 48] to the mother and another one to the young polyp. Each began to swallow its millepede. I saw the millepedes at first enter gradually, one into the stomach of the young polyp, and the other into the portion of the mother's stomach extending from her head to the spot from which the young one protruded. When the millepedes reached the junction of the two stomachs, they both passed into the shared portion, that is, into the posterior part of the mother's stomach. Nothing remained in the separate sections, either in the young polyp's stomach or in the portion of the mother's stomach reaching from her head to the emerging offspring. It was in the shared portion of the mother's stomach that the two millepedes were digested and reduced to a kind of pulp. This material then began to be tossed about noticeably in the mother's body; it came and went from the shared stomach into the separate sections and back again. Sometimes all the material passed into one of the separate sections and at other times both sections received a share. I have repeated this experiment very often with all the variations to which it lends itself.

Having witnesed the facts I have just reported, I could no longer harbor the least doubt in the world that a communicating opening existed between a mother polyp and her young one. It was clear that the skin of the young polyp was absolutely nothing more than a continuation of that of its mother, and that the polyps, like many plants, genuinely multiply by giving off shoots.

Further, as I have already mentioned several times, the young polyps do not sprout at one single fixed location. They emerge successively and often concurrently at many spots. I even venture to assert that there are few or no places on the bodies of some polyps which I have maintained for over two years from which young have not emerged.

Long-armed polyps do not grow young on their constricted posterior portion which forms a sort of a tail (Plate 8, Fig. 8, *b c*). As for polyps of the second species, however, I have seen them sprout young almost to the very tip of their posterior end, although in truth quite rarely.

The second experiment that I reported (p. 101) proving that the stomachs of young polyps communicate with those of their mothers, enabled me at the same time to learn the function of this connection. As we saw, the nutritious substance extracted from the food in the stomach of the mother polyp was carried into the stomach of the offspring. Serving this purpose is that peristaltic motion sometimes noticed in the polyps, the presence of which I substantiated in the preceding *Memoir* (p. 75). There I showed that this motion acts to spread the nutritious substance through the entire stomach and also to drive it into the arms and then send it back again into the body.

It was indeed likely that the food digested in the stomach of a mother polyp and then passed into the stomachs of her offspring served to nourish them. I sought conclusive proof of this fact, however, and it was not difficult to find.

I gave to various mother polyps some of the vividly colored food that I had used for the experiments reported in the second *Memoir,* that is, those worms that were flat and had viscera full of a red material (Plate 7, Fig. 7). A part of the red food passed into the stomach of the young polyps, which, like their mothers, took on a red color. This experiment proves that the food given to the mother served to nourish the attached young polyps, as can be concluded from the earlier discussion on the matter (*Memoir* II, p. 78 and following).

It is not surprising that an animal which remains joined to its mother should draw nourishment from her, as a great number of experiments have shown us for a long time. But polyps show us more than that; to wit, offspring which feed the mother while they are still joined to her, just as the mother feeds them.

As I have already mentioned, before these little ones separate from the mother they seize, swallow, and digest prey. After the food is digested, it then passes from the stomachs of the offspring into that of the mother in the same manner as the food taken in by the mother passes into the stomachs of her young. I have seen this happen very frequently. The reader may readily surmise that I did not fail to give red or black food to a number of offspring while they were joined to their mother. The result was as I had expected: the mothers took on the color of the food eaten by the young. They had been fed by their offspring.

When polyps are placed in locations abounding with the small creatures that serve them as food, often the mothers and the young independently devour several victims during the same period. These prey, which at first remain separate in the respective stomachs, become intermixed afterwards as digestion reduces them into a liquid or a pulp. This intermixing is easy to see, for example, if one gives a worm with red entrails to a mother and a piece of black slug to the young one. Some time later, a red substance can be found in the stomach of the young polyp and a black material in the stomach of the mother. I have even seen each of these substances, the red and the black, passing from one stomach into the other. At first they were quite separate; but, as a result of being driven back and forth from one stomach to the other, they mingled together and formed a single mass that was a color blended of red and black and that became the color of both the mother and the young one after digestion was completed. This experiment proves clearly that the mother and the young benefit in common from the food which each takes in separately.

I have shown that a young polyp feeds its mother. I should add further that it can also feed other offspring which the mother may be carrying at the same time. The nutritious substance which the young polyp has drawn from the food it has taken in, after passing from its stomach into the mother's, next spreads into the stomachs of the other offspring. The mother and her young can be thought of as a vessel having many branches freely communicating with each other. Thus, for example, when red food is given to one young polyp, the others soon are seen filling up with the substance extracted from it. Then, like the mother and the one which has eaten for them all, the other young take on a red tinge, the intensity of which varies with the amount of the nutritive substance.

As the posterior end by which a young polyp is joined to its mother constricts, the connecting opening between their stomachs narrows and finally closes. The communication then ceases. In order to determine how long the connection continues between the stomach of a mother and that of her offspring, I fed some polyps regularly, and each time I fed the mother or the young one I took care to examine attentively whether or not the substance was still passing from one stomach into the other. In the beginning, fragments of food passed back and forth between the stomach of the mother and that of her offspring. Then, the connecting opening having become smaller, nothing more than the nutritive juice could be transmitted without the fragments of food. Finally, while giving red food to the mother or the offspring, I noticed several times that no noticeable amount of juice passed from one to the other. Since the vividness of its color would have made the passage of the juice very easy to perceive, I concluded that the connecting opening, the aperture in the mother's skin, was then closed.

I have successfully performed the experiments just discussed on polyps of the second and third species.

Yet these experiments do not always succeed as one would wish. Polyps do not always accept the food offered to them with equal eagerness, nor are they always located so as to be conveniently observed through a magnifying glass. Thus, one should not become disheartened by want of success, but should try anew whatever has failed. It is even good to repeat successful experiments a number of times. All that it is possible to see is not discovered, and often cannot be discovered, the first time.

The second *Memoir* (p. 70) describes two polyps contending with each other for the same prey. Such competition occurs even between a mother and her offspring. Sometimes the mother takes away the water fleas or worms that her young one has started to swallow, and she partially ingests the young one itself as well. At times the little one also makes an effort to take away the prey which its mother has seized. I have never seen these struggles end in the death of either, however, and they finally part without having been harmed.

Until now I have made only the general comment that a polyp produces a number of offspring at the same time, but I have not yet indicated the extent of the fecundity of these animals.

In order to verify the fecundity of the polyps with one's own eyes, it is sufficient to nurture a few systematically during the summer. When I was raising quite a large number of them I took many precautions in carrying out my experiments, not only because I did not want to omit anything that might help throw some light on their fecundity, but also because the polyps I was using were to serve in other experiments requiring a great deal of precision. It therefore seems necessary to me to disclose here all the precautions that I took because the proofs of the facts which I shall be relating must be based upon them.

As subjects for these experiments, I chose young polyps newly separated from their mothers and placed each one alone in a separate powder jar. To avoid the risk of confusing the jars, I glued a number on each of them and noted the same number in the book that I had prepared for writing the journal of my observations. I always kept the jars containing the solitary polyps filled with suitable prey, using those water fleas discussed at length in the preceding *Memoir* (p. 58). No food is more convenient. A polyp in a jar well stocked with water fleas can continually eat as much as it wants, a necessary condition for the success of my observations. I had the good fortune to find all the water fleas I needed regularly for more than six weeks. Every two days I gave my polyps a fresh supply of water fleas along with fresh water, and I cleaned the jars when necessary. I always performed these tasks myself, because I wanted to be absolutely certain that each polyp had been replaced in its proper jar; I took all possible precautions to avoid any mistakes. Twice a day I visited my jars, carefully examined the polyp contained in each, and recorded in my journal what I observed.

Between the 1st and the 10th of June, 1741, I placed twenty polyps of the second species in solitude, and between the 20th and 23rd of the same month I did the same with fourteen polyps of the third species. Within a few days, three out of the twenty polyps of the second species, and eleven of the fourteen from the third species had died. I had isolated a good number of polyps so that my observations would not be interrupted by the death of some. Experience proved to me that this precaution was necessary.

I carefully recorded in my journal the order of birth of the offspring produced by my solitary polyps, including the day on which each young began to grow and the day on which it separated from the parent. As soon as I found a young polyp that had detached, I removed it from the jar holding the mother. I also isolated and observed with the same precautions those I had selected as specimens of their generation on which I wished to perform the same experiments as on the preceding generation. Accordingly, I placed alone in separate jars twenty young polyps of the second species and eight of the third, offspring of the first generation of solitary polyps. I continued to isolate them up to the sixth generation. The entry in my journal for each offspring included a record of the parent from which it had originated, the day on which it had begun to grow, and the day it had separated.

To illustrate my procedure, I will insert one section from my journal here. I have selected the record of a polyp of average fecundity compared with the many that I have observed. It is a polyp of the second species, a specimen from the third generation that had been raised in solitude. It began to emerge from its mother on the 5th of July and separated from her on the 10th; at that time it already had a young one that had begun to grow on the preceding day.

Order of young polyps	Day on which they began to grow	Day they separated from the mother	Order of young polyps	Day on which they began to grow	Day they separated from the mother
	July	July		August	August
1	9	13	24	14	17
2	11	15	25	14	18
3	12	17	26	15	20
4	13	17	27	16	20
5	13	22	28	17	21
6	16	22	29	18	22
7	17	23	30	18	23
8	18	24	31	20	23
9	22	25	32	20	24
10	22	26	33	21	25
11	23	26	34	21	26
12	25	27	35	23	27
13	26	29	36	23	28
14	27	30	37	24	28
15	29	—	38	26	30
16	30	—	39	27	31
	August	August			September
17	—	—	40	28	1
18	—	9*	41	29	2
19	—	11	42	29	2
20	9	13	43	31	4
21	11	14		September	September
22	12	15	44	1	—
23	13	17	45	4	—

*See Errata, Book I, p. 60.

I did not record the days on which the young ones 17, 18, and 19 began to grow, nor the days on which 15, 16, and 17 separated from the parent. Some distractions prevented me from making careful observations on those days.

The preceding table shows us that the polyp produced 45 young within two months, in other words, about 20 per month. As I mentioned before, among all the polyps that I nurtured and observed in the same manner, there were some which multiplied more and others less. I have found that the median rate between the two extremes was 20 young per month.

By examining this table it is easy to determine the approximate time that the young polyps remain joined to their mothers. One sees that it ordinarily lasts three or four days. Actually, however, the polyps reach the stage of separation that quickly only in the summer.

To acquire an idea of the fecundity of the polyps, it is not enough to know that a single individual can produce twenty young in a month. This fact would not be significant if the generations of their progeny did not succeed each other very rapidly. When I began to raise polyps in separate jars, I did not yet comprehend their fecundity. At that time it was the middle of June, and I feared I was starting too late to be able to observe at least a few successive generations. But my curiosity was soon satisfied beyond my hopes. At the end of a month, I already had noted in my journal observations spanning six successive generations of polyps. These observations taught me that four or five days after it has begun to emerge from its mother, a young polyp can begin to bear offspring itself; that is, a new generation of polyps can appear every four or five days.

Thus, in order to determine the number of polyps that can be produced by a single individual within, let us say, two months, one must first postulate that the original polyp can produce forty young during this period. In addition, however, one must calculate that these forty each begin to multiply five days after they have begun to grow, and that they also multiply at the rate of twenty per month. The same applies to the third and the following generations as well. Taking all of this into consideration, it is easy to understand that at the end of two months the number of descendants from a single polyp can be prodigious.

In the preceding discussion I have presupposed that all the polyps produced young and therefore that they were all mothers. I shall soon have the opportunity to prove it.

In performing the experiments just reported, I discovered a remarkable fact which I have already mentioned in passing and which I now need to discuss at greater length. While examining young polyps still attached to their mothers, I saw one that already had a little polyp starting to grow from its body. Hence, it was functioning as a mother while still united to its own mother. I soon found that what I marveled at in this polyp was common to many others. In a short time I had a number of young polyps still attached to their mothers; some of these young already had three or four offspring themselves. Several were even fully formed (Plate 8, Fig. 8), fishing for water fleas like the others and eating them. That is not the whole of it, for I have even seen a mother polyp carrying the third generation: from the young produced by the mother, another emerged, and from that one, a third.

When looking at a mother loaded with young (Fig. 8), one must be well convinced beforehand that this is an animal in the process of multiplying. One would more readily take it for a plant or for a root dividing into many branches.

I shall describe only one mother from among the great many that I have had the opportunity to observe. At the same time this illustration (Plate 8, Fig. 8) will furnish an example of the rapidity with which the polyps grow and multiply.

Our subject is a long-armed polyp. Fifteen days after beginning to emerge from its mother and nine days after separating from her, this polyp measured at least an inch and a quarter in length when well extended. Ten young were growing simultaneously from its body, and four or five of these were about sixteen to eighteen millimeters long. Eight of these young were fully formed and able to eat. Furthermore, five were producing little ones themselves. One of these five bore three offspring, two others each carried two, and the last two were sprouting one each. Some polyps of this second generation already had arms and were even catching water fleas. The entire grouping composed of the mother and the nineteen young was at least an inch and a quarter long and an inch wide. The arms of the mother, as well as those of the young, hung down for the most part toward the bottom of the powder jar.

Since this polyp had begun to multiply, only one offspring had separated from it. Thus at the end of two weeks, she was heavily laden with young. The polyp had always hung suspended at the surface of the water, and the jar in which it was kept had always been amply stocked with water fleas. These two circumstances, well suited to forestall too early a separation of the young from the mother, consequently resulted in the spectacle of this polyp so richly bedecked with young.

My remarks up to the present have made it easy to understand that the polyps multiply most when food is plentiful and the weather is warm. It is absolutely essential that these two circumstances occur together in order for the polyps to multiply as extensively as described in the foregoing discussion. Abundant food is no longer useful to these animals once the weather is not warm because, as I have shown in the preceding *Memoir* (p. 71 and following), the appetite of the polyp is proportional to the warmth of the surrounding water.

To be convinced that the polyps multiply in proportion to the quantity of food they consume, it suffices to make them alternately fast and eat for a period of time. Sometimes one performs this experiment unintentionally because animals suitable as food for the polyps can become scarce or even completely unavailable. I experienced this predicament especially at the start of my research when I was not yet adept at finding all sorts of food appropriate for the polyps. In times of scarcity of food for my polyps, they became smaller and ceased multiplying; but as soon as I supplied them with food, I had the pleasure of seeing them once again produce young.

I did not stop with this general kind of experiment. On July 18th, 1741 I isolated a polyp in order to deprive it of food. Four young were then emerging from its body, and yet a fifth one began to grow. The five young separated on July 26th. The mother remained without food until August 7th. During this time she bore no new offspring. In contrast, those polyps living midst an abundance of food sprouted new young daily. On the 7th of August I gave the mother some water fleas and on the 9th I saw a young one emerge; subsequently the mother produced a number of young. In short, I subjected the same polyp several more times to alternate periods of abundance and famine and always obtained the same result.

If one did not know that young polyps can secure their own food as soon as they have arms, although still united to the mother, it would be difficult to conceive how a polyp within two weeks could become so large and bear nineteen well-fed little ones. I am referring to the polyp described just previously (Plate 8, Fig. 8). This mother polyp consumed about ten water fleas each day, and certainly the nutritive substance drawn from those prey would not have been sufficient to feed all her offspring. From about the tenth or the twelfth day, however, those young which were capable of eating, daily devoured about twenty water fleas amongst them. This food blended with that of the mother, as I explained previously (p. 103), and the combined substance which resulted then dispersed into all the young polyps by means of the connecting openings between their stomachs.

Among the large number of polyps which I have fished from ditches at the most propitious season, I never found any that carried more than seven offspring at a time on its body, and even those I found only rarely. Apparently the creatures the polyps use for food are seldom as plentiful in the natural habitat of the polyp as they are in the powder jars in which I keep my polyps. I saw places in the ditches and in a fishpond, however, where for a long period of time the polyps could have seized as many prey as those that were in my jars. Once I had witnessed the prodigious fecundity of the polyps, I thought that it would take only a brief time for the ditches to fill with an infinite number of these animals even though but a few had been there shortly beforehand. For such multiplication to occur, the ditches would have to be well stocked with water fleas. This was only a conjecture, though certainly a highly probable one, that I formulated in August 1741. During July of the following year, however, experience substantiated it completely.

Since beginning to observe the polyps, from time to time I examine the waters of Sorgvliet in which I originally found them. I survey the edges of the ditches and of a fishpond, and from various places in the water I draw out plants or other objects on which I look for polyps. I also try to observe polyps in their natural habitat; toward that end I take the precautions indicated in the preceding *Memoir*. When I visited the waters of Sorgvliet in the beginning of July, 1742, I found very few polyps in the ditch located at the foot of a dune that had furnished me the first polyps I had ever seen. About two weeks later I was inspecting the edges of the ditch at a time when the sun, shining into the water, illuminated certain areas all the way to the bottom. All at once I saw a plant more heavily laden with polyps than any that I had ever seen. Eagerly I pulled it from the water. Regarding it as a treasure, I abandoned all else to put it safely in my study. After admiring it with delight for a few moments, however, I returned to the edges of the ditch and continued my survey. How overjoyed I was to reach a spot where I saw the entire bottom, as it were, bristling with polyps! Branches of trees several feet long which had fallen into the water were, without exaggeration, almost as fully decked out with polyps as is a wig with hair. One plank about ten feet long which was floating on the water was so completely lined with polyps that one would have described it as encompassed by a fringe.

I drew from the ditch one of those branches that was thickly covered with polyps. Once it was out of the water, all those polyps contracted, piled up one on top of the other, and formed a thick brownish slime which was not readily recognizable as a mass of polyps. I placed a piece of this branch into a large jar filled with water; a number of people observed it. The polyps were so densely packed that it was difficult to see the wood on which they were fastened. When their arms were extended they occupied almost the entire jar. An idea of the effect produced by the polyps and their arms can be gained from the figure in Plate 9. I must caution that however great the number of polyps shown here fastened to the wood, this number nonetheless does not come near the quantities I have seen on similar and on much larger pieces of wood.

As a diversion I threw some worms into the midst of the prodigious number of arms filling the jar that contained this multitude of polyps. A single worm was seized by a number of polyps, and in a little while it was entwined in their arms in myriad ways. However entangled the arms of the polyps holding a single prey became, I noticed that they subsequently separated themselves. For a number of days I watched this multitude of fine threads which almost touched one another and did not see any disorder among them (Plate 9). They moved in various ways, elongated, and contracted without becoming tangled together.

After carefully examining everything within eyesight in the afore-mentioned ditch, I realized that the greatest concentration of polyps occurred on the branches and twigs lying closest to the surface of the water. The idea came to me that the majority of the polyps were very likely drawn there by their propensity for light; I sought at once to verify this supposition by an experiment. I implanted some sticks in the bottom of the ditch, arranging them in such a way that the top ends reached almost to the surface of the water. My aim was to see whether the polyps which were so numerous on the bottom of the ditch would ascend along these sticks and proceed to locate themselves in large numbers at the top ends where there was the greatest exposure to the light. A few days after I had arranged the sticks as described, however, the number of polyps in the ditch decreased quite suddenly, as I shall soon discuss, and consequently it was impossible for me to see the outcome of the experiment that I had started.

I found it very easy to determine why there was such a prodigious quantity of polyps in that spot. The ditch was richly stocked with water fleas which had provided all the food required for the rapid multiplication of both the original polyps and subsequent generations of their descendants.

I took advantage of the brief period when the ditch abounded with polyps to observe as fully as I was able all their maneuvers in their natural habitat. Each day I went to lie on a plank which projected a little above the water.

Keeping my eyes fixed close to the surface, I saw the polyps walk, extend their arms, and seize the animals that touched them. When the wind ruffled the water, I saw large numbers of polyps suspended at the surface move with the waves. Bands of polyps, carried along by the movement of the water, passed under my gaze.

Most of the snails in the ditch at that time bore polyps upon their shells. These polyps had no need to walk in order to journey from one location to another. The snails served as carriages for them, and although they advanced very slowly, nevertheless they allowed the polyps to travel further in a short time than they could have done in a day of walking. Other polyps went still faster. They were attached to the cases of the pretty caddis worms, cases composed of small pieces of plants cut into long rectangles and joined together in a spiral shape. These caddis worms swim rather rapidly. I have seen some swimming along carrying a number of polyps fastened to their cases (Plate 10, Fig.1).

I have already said that this great abundance of polyps did not last for long. A week after having noticed the pieces of wood laden with polyps, I found nothing but their remains; these can be recognized only by those who have previously seen dead polyps. It is highly probable that the lice I described earlier [p. 82] contributed greatly to the destruction of these polyps. All the polyps that I took out of the water at that time were infested with lice. When I left the polyps in large jars without ridding them of the pests, most were devoured by the lice.

As I stated earlier, polyps are able to multiply during rather cold weather. During November, December, and March, I found in the waters of Sorgvliet some which bore three and four young; at the same time, the polyps in my study were producing even more. The experiments I carried out proved clearly to me that the polyps can produce young throughout the whole year. A number of the polyps I have maintained for a long time can serve as examples.

As indicated in the entry from my journal (*Memoir* III, p. 106), the polyp under consideration began to sprout young on the 9th of July, 1741. From that time to the present (January, 1744), it has never ceased to have some offspring attached to its body, even though during the winter I kept it in a study where it was but a few degrees warmer than outdoors. I removed it to place it on the mantel of my fireplace only in freezing weather.

Having recorded the dates on which the polyps were born during the summer, I was eager to keep a record during the winter. I started this experiment in the autumn of 1742, choosing for subjects two polyps of the second species and two of the long-armed species. My purpose was not limited to investigating how many young they would produce in one month of winter; I also would try to observe as precisely as I could the degree of cold at which they would cease to multiply and at what temperature they would start reproducing again. Further, because the quantity of food they consume has a great influence on the number of young that the polyps produce, I intended to note what they ate in winter and to make them eat as much as they could. By these means I hoped to acquire more complete ideas about all the factors affecting the reproduction of the polyp.

Since these experiments required that I constantly note the temperature, I placed a mercury thermometer of Mr. Prins in one of the four jars containing the polyps I was observing; I kept the other three jars close by it. We know that a reading of 32 degrees marks the freezing point; 48 degrees, temperate air; 64 degrees, warmth; and 212 degrees, boiling water. According to the observations that were made by Abbé Nollet comparing the scales of the Réaumur and Prins thermometers, ten degrees on the scale of Mr. Réaumur equal twenty and two third degrees on the scale of Mr. Prins. Those who wish to render my observations according to the Réaumur thermometer may do so with ease by guiding themselves according to this ratio.

Ordinarily I scanned the thermometer three times daily: in the morning, at noon, and in the evening. Each time I noted relevant observations in a record book. It does not seem necessary to insert this record here. It will suffice to indicate the following conclusions which can be drawn from those observations that I recorded:

1. Polyps are able to eat when the temperature of the water surrounding them drops to thirty-five degrees on the Prins thermometer, although in fact they often refuse to eat when the water is that cold. When the temperature rises close to forty degrees, they can consume a meal every two or three days; but this meal is not half as large as those they eat in the summer. They need two or three days to digest the food they have eaten.

2. Some offspring began to grow when the thermometer read thirty-eight degrees, but young sprouted more frequently when it rose to between forty and forty-six. Such a temperature is not sufficiently warm to cause the sprouting of most plants that we cultivate in our gardens.

3. It further appears from my record book that in cold weather the majority of young polyps remain attached to their mothers for about a month. They do so not because they need all that time to attain sufficient

maturity to separate, but because they are less lively at those temperatures. Not distressed by hunger, they are not stimulated to change their location in order to seek food.

Before making all the observations just reported concerning the fecundity of the polyps and how it is influenced by the amount of food they consume and by the temperature of their habitat, I had already tried to discover how they were fertilized.

I continued these investigations while conducting the studies on fecundity. My findings on reproduction in polyps had shown it to be more like that known in many plants than in a great many animals. Thus, in attempting to spy on the love life of the polyps, I had in mind not only possible resemblances to animal reproduction, but to that of plants as well.

First I confirmed the fact that all polyps with arms shaped like horns are mothers, that every individual of the species bears young. Most of the polyps which I first encountered in the ditches were not producing young. However, after I had them eat some meals in the powder jars where I put a great number of them, every single one began to multiply. This experiment made me conjecture that perhaps all polyps were mothers. To reach greater certainty in deciding the matter, I devoted a great deal of attention to the large number of polyps I was raising and using as subjects for various experiments. Every single one of those polyps, in fact, began to sprout young once it had consumed some quantity of food. From that time on I had no further reason to doubt the correctness of my conjecture.

Since then I have raised, I venture to say, thousands of polyps, and I have not found any that failed to reproduce after consuming a certain amount of food.

As I scrutinized the polyps that I was keeping in groups in the jars, I watched their every move all the while in order to note whether there transpired between them anything analogous to the process of copulation found in most animals. No matter how carefully I tried and how large the number of polyps I observed, however, I never succeeded in detecting anything of the kind. Thus, I wondered if perhaps the polyps resembled aphids which, according to the recent discovery I described on page 10, were all mothers and multiplied without mating. Such a model, however, was not sufficient. Evidence would have to be found and confirmed by experiments performed on the polyps themselves.

Thus, I isolated a number of polyps of the second and third species. I wished to select individuals I was quite certain could not have communicated with other polyps since separating from their mothers. Thus, I chose only two sets of polyps. The first were those which I myself had separated from the mother. The second were those which I had removed from the jar immediately after they had separated from the mother on their own, but before any other offspring had separated with which they could possibly have mated. Earlier, on page 105, I described all the precautions I took to rear these polyps in complete isolation and to insure that I could always identify them and ascertain that they and their descendants, isolated in a similar manner, had all multiplied.

In fact, all those polyps which lived alone for some days and which took food, produced young ones and continued to produce more and more of them in proportion as I fed them.

Not only did these polyps that I isolated multiply, but so did numbers of their descendants, up to the seventh generation, which I also kept isolated using the same precautions.

I have performed this experiment with a considerable number of polyps. Thus one may conclude quite confidently that after a polyp has separated from the mother it need not have any communication with another polyp in order for it to multiply. I became convinced of this particular fact shortly after beginning to raise isolated polyps. I saw a number of instances of young that were still attached which were themselves starting to produce offspring (p. 108). To state the matter in different terms, polyps possess the ability to reproduce even before they detach from the mother.

After I had become convinced that the young polyp, once separated from its mother, had no need to live with others in order to multiply, and after I had seen polyps multiply while still attached to the parent, I thought that perhaps something occurred during the union of mother and offspring that contributed to the fecundity of these animals. Consequently, with great care, I set about systematically observing young polyps not yet separated from their mothers. Despite all the pains I took, however, I could never detect anything either between the young attached to the same mother nor between these young and their mother, on which to base even the slightest suspicion. All these polyps were multiplying without my having been able to note anything which could have contributed to their reproduction. I therefore concluded that what I was attempting to discover, supposing something of the kind even existed, was either imperceptible or at the least very difficult to see.

Thus I was reduced to groping my way along even more hesitantly than before.

Since the process by which plants are fertilized is well hidden, even invisible for the most part, it occurred to me that perhaps the polyps resembled plants yet again in this respect. It was very difficult, not to say impossible, to be able to determine whether such a resemblance did in fact exist between polyps and plants. For this reason I considered the idea to be merely a most dubious conjecture which did not merit much attention. Nevertheless, it seemed appropriate that I not neglect to perform an experiment which this conjecture had brought to mind.

A pistil and stamens are found in the flowers of plants; I wanted to learn if there were any parts at all performing the same functions in the polyps.

Regardless of whether fertilization was effected in the manner of plants or of animals, or even in a manner completely unknown, I predicated that perhaps these supposed reproductive parts could be located in the head and arms of the polyps. I set about investigating first if mothers and their young did not in some way impart the source of fertility to each other by means of their heads and arms. With this aim in view, I severed the heads of eight mothers bearing offspring which had not yet sprouted arms. I isolated each of the eight mothers in separate glasses. As their arms and heads grew again, I cut them off again. Thus they had neither arms nor head during the entire time that the young remained attached to their bodies. Consequently, no communication by means of the arms and heads could have occurred between these little ones and their mothers to contribute to fertilization. As the young became capable of taking food, I fed them. They grew and separated from the mother. Immediately after they separated, I isolated them in other glasses. Had it been necessary that the mother impart a principle of fecundity to them in the manner I had conjectured, they should not have been able to multiply. After a few feedings, however, all multiplied prolifically, proving that conjecture to have been false.

Possibly, however, this supposed communication might take place between the young ones which emerged from the same mother at the same time. There was scarcely any likelihood that this idea was true; nonetheless, I did not wish to neglect to perform the experiment that could settle whether or not young polyps made any use at all of their heads to fertilize one another. Therefore I severed the head of a mother carrying only a single offspring; it remained alone the entire time until it separated. This young polyp multiplied prolifically while it was in the glass in which I isolated [and fed] it after it had separated from its mother.

I performed one other experiment which likewise proves that a young polyp has the ability to reproduce within itself before it could possibly receive the principle of fecundity externally from its mother or from any other polyp. I cut away a young polyp that had barely started to sprout, that is when it was still only a very small bud such as the one marked *e* in Figure 1 of Plate 8. Figure 9 of the same plate shows it as it appeared immediately after I had used a scissors to sever it from its mother. In Figure 10 it is drawn enlarged as seen under the microscope. This little bud placed alone in a separate jar gradually grew longer, developed arms, and then multiplied.

Finally, to conclude this discussion, let me state that however attentively I tried to see whether or not a young polyp fertilized itself in some external manner, I could detect nothing which gave the least hint of what I was seeking.

It seems advisable to recall here in brief what we have learned from the various experiments which I undertook to discover the principle of fecundity in the polyps. From them we can conclude the following:

1. That a young polyp, after it has separated from the mother, does not require the company of another polyp in order to multiply.

2. That a polyp has within itself the ability to reproduce even before it separates from the mother, because it multiplies during this period.

3. That if the mother polyp does transmit the principle of fecundity to her offspring while it is still joined to her, the transmission does not take place between the arms or head of the mother and the young.

4. That neither is a young polyp fertilized in this way by the head or arms of another young polyp emerging from the same mother at the same time.

5. That if a young polyp fertilizes itself, it is quite probable that the process is imperceptible.

Everything included in these five points applies to polyps of the second and the third species. The experiments which gave rise to these conclusions were performed on both. I have not been able to perform them on the green polyps because I lost them before advancing my investigations that far and have not been able to find any since.

If we have not been able to discover how the polyps are fertilized, we have learned at least that they definitely are not fertilized in the manner of most animals known to us. Thus, polyps again constitute an exception to an allegedly universal rule, that *there is no reproduction without copulation.* This rule had already been very strikingly contradicted by the discovery made a few years ago on aphids.

While I was making all the observations reported previously on the polyps' mode of multiplying by putting forth buds, I continued to investigate the possibility of their having other natural means of multiplying. For example, I took great care to note attentively whether or not they divided on their own, that is, whether these animals naturally multiplied by dividing or whether, to the contrary, this type of multiplication occurred only when one cut them into two or more parts. As a matter of fact, I have seen polyps divide on their own. The area where they were to divide gradually constricted until finally a small movement could effect the separation. Each part then became a complete polyp through the same process of reproduction as that occurring in each half of a polyp which has been cut in two. Regarding the hind parts of these polyps which divided on their own, I observed that they required from two to three weeks, even in summer, to regenerate into complete polyps. Some polyps divided themselves at the middle of the body and others more or less near the anterior or near the posterior end. Some polyps that I maintained for a long period divided by themselves two or three times. The shortest interval between one division and the next lasted three months. Two of these polyps came from parts of one of those animals that I had sectioned successively into fifty pieces.

Although I have studied a considerable number of polyps over a period of nearly three years, I have not seen more than twelve divide on their own. If this kind of multiplication does not occur more frequently in the polyps' natural habitat, I do not know whether it can be regarded as a method that Nature has given the polyps to perpetuate their species, or whether it should be considered rather as an unusual occurrence. Whatever the answer, it appears that they seldom multiply by this means, and that it is in no way comparable to their multiplication by budding.

I also tried to learn whether polyps were oviparous. I have never seen anything that could be taken for eggs, except what I shall now describe:

I have noticed small spherical protuberances fastened to the bodies of some polyps by a very short stem (Plate 10, Fig. 2, *e, e*). Some were white, some yellow. I never saw more than three at a time on an individual polyp. After remaining attached to the polyp for some time, they separated and fell to the bottom of the glass. I examined them on various occasions before and after they separated. In the end they all came to naught with the exception of one which possibly became a polyp. I say possibly because I am not altogether sure. At the time when I should have been observing this protuberance most attentively, two days went by without my looking at it. When I came back to examine it and looked in the place where I had left it, I found there an imperfectly formed polyp which appeared in fact to be coming out of a spherical object on the bottom of the container. The side of the sphere which touched the bottom was elongated. On the opposite side, which was still rounded, the tips of three arms could be seen starting to emerge. Gradually, this polyp elongated and assumed the usual shape of these animals. In short, I would have been certain that this polyp came from one of those small spherical bodies which had detached from a polyp had I not missed two days of observing it and had there not been some other little polyps in the same container.

Mr. Allamand, who has also observed a number of these small bodies, has had an almost identical experience. It seemed to him that one of them became a polyp. Nonetheless, he does not venture to assert it positively, because the experiment was not carried further with sufficient exactness to satisfy him.

Thus it is advisable that these observations be repeated in order to determine more precisely what becomes of the small spherical objects under discussion. Assuming that they become polyps, however, I do not know whether they should be viewed as eggs or as polyps which accidentally take on a spherical shape while growing and then either die or develop into complete polyps.

Mr. Réaumur honored me with a letter written on December 17, 1743, in which he informed me that Mr. Bernard de Jussieu, during his vacation, had found a large number of polyps with arms shaped like horns which bore small vesicles upon their bodies. It seems to me quite probable that these small vesicles are the same things I have called small spherical bodies (Plate 10, Fig. 2, e, e). They appeared to Mr. de Jussieu to be full of eggs; but as he was obliged to continue his journey, he was unable to ascertain satisfactorily what those eggs yielded. If they were indeed the eggs of the polyps with arms shaped like horns, these animals would be both oviparous and viviparous. In truth, this statement is as yet mere conjecture, but a conjecture formulated by a naturalist such as Mr. de Jussieu is worthy of the greatest attention.

In addition I should caution that I have found the spherical bodies solely on polyps of the second species and only during autumn and at the beginning of winter.

Also in autumn and at the beginning of winter one discovers on polyps of the second and the third species small protuberances quite different from those under consideration until now. They are irregular in shape, some being almost pointed, while others are either flattened or rounded on top (Plate 10,

Fig. 4). Instead of holding to the polyp by a stem, as do the spherical bodies described previously, they are fastened to it at their widest portion. Their form is nearly that of a pyramid, the base of which is pressed directly against the polyp. These protuberances on an individual polyp are at times so numerous that they nearly touch. This abundance of protuberances is noticeable principally on the polyps with long arms which carry the protuberances only on the widest portion of their bodies, that is the area from the head to the beginning of the tail (Fig. 4, *a c*). All these protuberances are white and easily noticeable because of their color, especially when they are present in large numbers. One might be inclined to regard them as a sickness afflicting the polyps, or at least as an aftereffect of a sickness. It is certain that these animals are not so active when their bodies are covered with these outgrowths; they show little or no appetite, grow thin in a short time, and lose their color. When polyps bearing the protuberances begin to eat again, the excrescences gradually become smaller and finally disappear completely, in contrast to the spherical bodies which separate from the polyps. [Note: Today the pyramidal protuberances are known to be spermaries.]

I have noticed that polyps of the third species which are heavily indisposed with these pyramid-shaped outgrowths have very thin and very short arms (Plate 10, Fig. 4). I have been unable to see them elongate.

Although they had eaten very little, the long-armed polyps that were covered with this latter type of protuberances produced a large number of extremely small offspring in a short time on a number of occasions during the winter. I counted twenty-two of these small offspring at the same time on the body of a single polyp. The young then separated from the mother, and those that were fed became as large as those growing during summer. At first I had suspected that these little polyps emerged from the protuberances, or rather that these protuberances became little polyps. When I examined the phenomenon more closely, however, I noticed that the polyps grew alongside the protuberances in the spaces remaining between them.

At present I shall discuss some other irregular growths that are seen in polyps when they are observed systematically for some time.

Ordinarily the arms of the polyps are undivided without any branches. Rather frequently, however, one sees a polyp that has a forked arm or two, and sometimes they fork in even more than one place. The branch which emerges from the arm to form the fork sometimes starts close to the base and sometimes closer to the tip of the arm. I shall describe here, exactly as I saw it, a particular arm of one polyp of the third species (Plate 8, Fig. 11).

The polyp had a very remarkable forked arm ($a f f f f f$). The arm first divided into two branches at c. One of these branches then forked again in two places, that is both at d and at e, and the other branch forked in one place at i.

I have seen some green polyps and some polyps of the second species which had one, two, or three arms elsewhere than in their normal location (Plate 10, Fig. 3, $c d$). They emerged from the polyp's body closer either to the anterior or to the posterior end. I have never seen more than three such displaced arms on one polyp except on those upon which I had performed some operation, an example of which will be shown in the following *Memoir* [p. 162]. I have not seen that the arms located elsewhere than around the mouth served any useful purpose for the polyps. Most of them gradually diminished in size and finally disappeared entirely.

In addition, I have seen on polyps of the second species some offspring that always retained the shape of a cone, and that had only a single arm which emerged precisely from the tip of the cone (Plate 10, Fig. 6, c, c, c). I do not know how these polyps come to be, for I noticed them only when they were already formed. It seemed to me that some of them, instead of separating from the mother, gradually diminished so that the little cone disappeared and in its place only a single arm remained. Others, to the contrary, first lost their arm and then separated from the polyp to which they were attached. I have never seen more than three conical polyps on the body of another (Plate 10, Fig. 6). It is only in autumn and in winter that I have seen them.

Sometimes one encounters polyps which seem to have two heads. I studied some of them when they were developing, that is when they were emerging from the body of their mother. When the arms began to grow, two heads instead of one appeared at the anterior end of these young polyps. At first the heads were contiguous, but next, after the polyp had developed somewhat, each of the heads elongated and formed two branches that were joined to the rest of the young polyp's body which they shared in common. I have kept such two-headed polyps for quite a long time after they detached from their mother. It was natural to mistake one of the two heads for an offspring which was beginning to grow at the same time as the arms of its parent. Under this assumption, the two young should have detached at the end of a certain period which would vary in length according to the season. In many cases this separation did not occur. Some of the young remained attached to their parents for a very long time. The spot at which they were fastened did not constrict nor did communication from one branch to the other cease. I have seen some, however, which finally did detach. Thus, the second head of a double-headed polyp can be regarded either as an extraordinary young polyp which has remained attached to the mother longer than the others, or as a second head of a polyp which has split into two portions on its own, as sometimes happens at the anterior end of a polyp.

Among a number of polyps which appeared at the outset with two heads, I observed one for a long time that I believe I should describe here. This was a polyp of the long-armed species. When it detached from its mother, I placed it in a shallow glass and fed it well. In a short while the two branches, or anterior ends of this polyp, lengthened considerably. At the same time the tail end which they shared gradually diminished until it finally disappeared. At first the two anterior ends were at an angle to each other; but after the common tail had disappeared, they were joined together end to end forming a polyp with a head at each extremity (Plate 10, Fig. 5, *a* and *c*). I provided it with food sometimes at one mouth and sometimes at the other; but regardless of the mouth that swallowed it, the food passed into the entire body. The middle section of this polyp (Fig. 5, *e i*) was narrower than the two ends (*a i, c e*), being of the same width as the tail of an ordinary long-armed polyp. This particular polyp sprouted young on both sides away from the narrowed middle portion. I watched this polyp very attentively to see whether it would walk, and if so, in what manner. During the two months that it remained in this peculiar state, however, I never once noticed it change its place. Every time I observed it I always found it lying on the bottom of the vessel in precisely the same spot where I had left it. Most of the time it was rather elongated, its arms winding about in various ways on the bottom of the glass (Plate 10, Fig. 5). It remained in this particular state that I have just described for two months. Finally, in its middle, it formed a very short tail once more (Fig. 5, *m*). The connection between the stomachs of the two branches closed, and each branch had the appearance of a complete polyp. It is evident that these two polyps would have separated, but they died of a sickness a short time after these last changes occurred.

If one sees polyps which sprout two heads, one also sometimes has occasion to observe others which do not grow a head at all. In such a polyp, no arms emerge at the anterior extremity, nor does it form a mouth there. These incomplete young polyps could be viewed as extraordinary tails which form upon the mother polyp; as a matter of fact, they function as tails. Often the polyp from which they grow uses them in the same manner as it uses its posterior end, that is to cling to objects both while moving and at rest. On January 18, 1743 I saw as many as seven of these extraordinary tails on a single polyp (Plate 10, Fig. 7, *q, q, q, q, q, q, q*), which, in addition, bore two ordinary young polyps and three of those conical polyps (*c, c, c*) which I discussed on page 121. The very accurate representation of it in Figure 7 shows the bizarre shape of that polyp. The polyp at that time was a year and a half old. By the following summer none of those extraordinary young remained on its body; it had resumed the appearance of an ordinary polyp.

Were I to describe in polyps that I maintained for a long time all the varieties of shapes produced by the irregular growths to which these animals are subject, I would have to enter into too much detail. What I have said about the matter is enough to make it clear that in regards to these irregular growths, the polyps yet again resemble plants more than they do the animals we know.

Having disclosed everything that I have been able to learn about the manner by which polyps multiply, and having shown that it closely resembles the processes familiar to us in plants, perhaps it would be useful now to examine the elements of this similarity in greater detail.

Plants multiply in three different ways: by seeds, by cuttings, and by shoots. All three types of multiplication do not occur at the same time.

Until now we have discovered nothing in polyps that we can identify with assurance as the seed; thus in this respect we cannot trace any similarity between them and the plants. On the other hand, however, we find that they are similar in regard to the two other processes. My remarks at the beginning of the first *Memoir* (p. 8 and following) on the reproduction that occurs in the parts of a polyp which has been cut into two suffice to make us realize that these parts greatly resemble the cuttings of plants. And in the present *Memoir* all that I have reported on the polyps' natural mode of multiplication has shown that the young polyps are in truth shoots which emerge from the old polyps, as suckers issue from a plant. The fact is, however, that the shoots of the polyps separate by themselves whereas those of plants must be separated artificially. At least this practice is necessary in gardens and nurseries where plants are made to multiply by shoots.

There is, however, one quite common plant from which the shoots separate by themselves. I am referring to Duckweed [*Lemna;* see Book I, p. 48]. This plant is composed of a single leaf which ordinarily floats on the water and which sends forth, from the middle of its lower surface, one or more rather fine threads that can be considered its roots. If during the summer one observes this plant with any regularity at all, he will see other leaves emerging in various spots from its edges and growing their own roots once they reach a certain size. When they are formed, these leaves or, to put it otherwise, these shoots, adhere very lightly to the plant from which they sprout. They then separate from it by themselves, and floating freely with the movement of the water, they can soon be found quite a distance away from the plant which produced them.

Thus it is established that polyps, like many kinds of plants, multiply both by being made into cuttings and by shoots, and that as a consequence there is a great deal of similarity between them in their mode of multiplying. This being the case, would there not also exist a similarity between the way in which plants and polyps are fertilized? All that we know about a large number of plants is that their flowers contain the parts which contribute to their fertilization, namely the pistil and the stamens, so that when these parts are missing, the plants are incapable of producing fertile seeds. At the present time this knowledge does not help us to understand fertilization in the polyps, because we have found nothing in them analogous to the flowers and seeds of plants.

The question arises whether or not the processes taking place in the flower influence the ability to reproduce by cuttings or shoots. For example, can the vine multiply by cuttings and the elm by shoots independently of what occurs in the flower? Or can this type of reproduction occur only when the process which renders seeds fertile has taken place in the flowers? To answer this question it will be necesary to enter into a short discussion. My remarks on the subject perhaps will help us to catch at least a glimpse of an even greater similarity between the polyps and those plants which multiply by cuttings and by shoots.

It is highly probable that the fertilization occurring in flowers bears only on the fertility of seeds and not on the reproduction by cuttings and by shoots. I base this statement on the very familiar fact that some flowers which are naturally single at times become double. We know that the plants which have single flowers produce seed, but those of the same species which have double flowers do not. We are aware that the plants with double flowers have been rendered incapable of bearing seed because the parts that serve to fertilize and form the seed have become petals. Thus the process that takes place in flowers and fertilizes the plant through the seeds does not occur in these plants with double flowers. It does not occur, for example, in a double wallflower plant. Nonetheless, independently of any such process, this plant multiplies by shoots. These shoots are able to produce others which, from generation to generation, will always be equally capable of multiplying in this manner. Thus what we learn from experience leads us to conclude that a double wallflower can multiply by shoots independently of what occurs in its flowers.

Perhaps one could extend this conclusion to include plants which have complete flowers and can multiply by seed. According to this reasoning, it could be said, for example, that the process occurring in the flowers of an elm which results in the fertilization of its seeds has no influence at all on the ability of that tree to reproduce by shoots. The tree can produce shoots independently of what takes place in its flowers, and furthermore the shoots it produces will themselves be able to propagate not only by shoots but also by seeds. Thus, as far as multiplication by shoots is concerned, it appears that plants are able to reproduce by themselves.

What I have said about shoots applies to cuttings as well. It is evident that whatever enables a cutting to become a complete plant and therefore to multiply subsequently both by cuttings and by shoots has no relation to the process within the flowers which affects the seeds. A runner from a vine, for example, would furnish cuttings equally well had it borne flowers or had it borne none of them at all.

Thus, could not these facts regarding plants that multiply by shoots and by cuttings hold true for polyps as well? Could they not be fecund in themselves, capable of reproducing by both shoots and cuttings, without there occurring in them any process analogous to those we know in many animals, or have reason to suspect occurs in the flowers of plants?

We are now in a position to conclude that the extraordinary manner by which polyps with arms shaped like horns can be multiplied through sectioning them into parts, which for so long a time was not known to occur in this class of organisms, is not the sole peculiar characteristic that distinguishes them from so many other animals. Another characteristic that must be included in the same category of exceptions is their natural manner of multiplying by shoots. Even should no other exceptions be found than those furnished by the polyps whose natural history I offer, their natural mode of reproducing by shoots can teach us twice over that the so-called general rules, which are almost universally accepted, do not deserve that name.

In the preface to the sixth volume of his *Mémoires sur l'Histoire des Insectes* (p. 55 and following), Mr. Réaumur states that a number of inquisitive persons, on learning that one could multiply the polyps by sectioning them, discovered the same property in various species of worms. Since that preface was written, this experiment has succeeded with an even larger number of animals. It is highly probable, as this famous naturalist states, "that the polyps are not likely to be the only animals which have been accorded the capability of being reproduced in so unusual a manner. The more one examines the products and the processes of Nature, the more one is convinced that none are unique." I believe that this statement applies to both methods by which the polyps reproduce. After having seen the buds develop on the polyps, one would have reason to think, in conformity with this principle, that however strange it seemed, this method of reproduction was not peculiar to these animals alone, but that it was, on the contrary, common to a number of other animals of different species and even of different genera.

However well-founded this conjecture might be, one would certainly wish to see it confirmed by examples. I did not need to wait a long time to find a most remarkable one. The appearance of the animals which furnished me with this example led me to classify them among the polyps. Their arms, located at the anterior end around their mouth, are arranged quite symmetrically and form a pretty tuft (Plate 10, Fig. 8, *a c d d d e*). The one depicted in Figure 8 is enlarged a great deal; in Figure 9 they are shown in their natural size at *p, p,* and *p.* In order to distinguish these animals from other polyps, I called them tufted polyps [the bryozoan *Lophopus;* see Book I, p. 48].

As I have already said, I found my first specimens of the second species of polyps with arms shaped like horns in April, 1741 while searching for green polyps. At the same time I discovered the tufted polyps. There were a number of them on the aquatic plants I had assembled in glass jars filled with water. These tufted polyps greatly excited my curiosity. At first they (Fig. 8, *i b a d d d e*) evoked in my mind the image of a blooming flower, and because there were a number of them together, they formed a kind of bouquet (Fig. 9). Now I need to describe the structure of these animals in a little more detail so that I can make myself better understood when I proceed to discuss the way they multiply by shoots.

The body of the tufted polyp (Fig. 8, *i b a e* and Fig. 9, *i e*) measures about two millimeters in length without counting the tuft (Fig. 8, *a c d d d e* and Fig. 9, *e*) which is almost as long as the body. The body is very slender and nearly cylindrical, and the skin is completely transparent. The tuft is but a continuation of this transparent skin; it is quite large in proportion to the body and is of a most remarkable shape. The base of the tuft (Plate 10, Fig. 8, *e a c*) is in the form of a horseshoe, and from its edges the polyp's arms arise (Fig. 8, *a d, a d, a d*), all curving gently outward. The tuft formed by this cluster of arms looks like the open blossom of a gamopetalous flower. The arms are very close together; I have counted more than sixty of them in a single tuft. They can be compared in their fineness and transparency to very thin glass filaments.

The base of the tuft is furrowed like a gutter (Fig. 8, *e*). At the middle part of its horseshoe shape, the base is attached to the polyp; at that place there is an opening serving as the mouth of the animal. The viscera are easily distinguished through the transparent skin of the body. In polyps that have eaten well, they appear rather dark brown. After having observed the tufted polyps for some time and having succeeded in seeing them eat, I was able to distinguish three principal parts in their viscera, namely the esophagus *e h,* the stomach *f g,* and the rectum *f a.*

It seems to me that these animals resemble plants even more strongly than do the polyps with arms shaped like horns. Accordingly, some very competent naturalists who have noticed different kinds of tufted polyps on the surface of various so-called marine plants have mistaken them for flowers and roots. Had they enjoyed the opportunity to observe in these tufted marine polyps a characteristic which they apparently share with our freshwater types, they would have recognized them as the animals they were. They are voracious, indeed extremely voracious; at least such is the case with those of the species that I have observed. It is true that they can eat only very small animals, but in a day they devour a great number of them.

For these tiny prey, the tuft of the polyp is an abyss into which most of them are hurled if they approach it while swimming. If one places tufted polyps in water well stocked with very small creatures and observes them attentively through a magnifying glass, he will notice quite easily the method by which they attract prey and make them fall into their mouths. From moment to moment an arm or two will suddenly bend towards the inside of the tuft and then return to its prior position. Rarely does the same arm bend twice in succession. If while examining the action of the arms, one casts a glance around the tuft, he will notice that, as the arms bend, the little animals swimming above are hurled into the tuft one after the other. The arms do not touch the prey at all, but they generate a kind of whirlpool in the water which carries the victim into the tuft. It frequently attempts to escape, but the sudden inflection of an arm gives a new degree of speed to the current dragging the victim along, and despite its struggles the animal is carried to the bottom of the tuft. I have stated that the bottom, or base, of the tuft is hollowed into a gutter. The little creatures destined to serve as the prey of the polyp thus fall into this gutter and then slide into the mouth located in the center (Plate 10, Fig. 8, *e*).

When the polyp is viewed from the side, it is easy to see it swallow its prey. It passes from the esophagus into the stomach and, unless it is extremely small, it remains distinguishable even inside the stomach, because all parts of these polyps are transparent. I call the esophagus that short canal *e h* extending from the mouth to a sac *f g* that serves as the stomach of the polyp. The esophagus ends a little below the upper end of the stomach at *h*. The food consumed makes the stomach clearly recognizable. As can easily be seen, the food is tossed about inside much more rapidly than it is in the polyps with arms shaped like horns. In the tufted polyps, the food is pushed back and forth alternately from the bottom to the top and from the top to the bottom. One can easily be misled about the actual length of the stomach and imagine that it extends up to the base of the tuft. To avoid the error, however, one need only note attentively the point to which the food is carried as it is pushed toward the top of the stomach. It will be seen to stop a little above the juncture of the esophagus and the stomach, and at that point (Plate 10, Fig. 8, *f*) it proceeds to return toward the lower end *g*.

There is an area between the upper end of the stomach and the base of the tuft that is occupied by a small sac *f a*; very frequently it is completely filled with a brown material of a deeper hue than that material found in the stomach. This sac is the rectum, and the brown matter is excrement. The excrement forms a roughly oblong and readily observable pellet which occupies all the space in the rectum. The rectum empties completely at one time, with the grain of brown matter that fills it coming out in one piece through an opening at the base or at the side of the base of the tuft. I have often seen these polyps void their excrements, but I have never been able to discover the exact location of the opening through which they are emitted. After the polyp has emitted its excrement, the empty rectum appears to be of a very light brown color and even somewhat transparent. If this animal is in water well supplied with prey and if it devours many, as it ordinarily does, fresh excrement will soon pass from the stomach into the rectum which will become filled and again take on the color and opacity it previously had.

I stated above that in order to draw prey into its mouth, the tufted polyp bends several of its arms inward. Sometimes the arms can be seen performing a different maneuver, as happens when an animal too large to be swallowed or some other object has fallen into the tuft. In order to rid itself of the unwanted matter, the polyp opens its tuft either completely or partially by spreading its arms out wide and then returning them to their regular position. In bending outwards and turning in again, all arms act in unison.

Polyps with arms shaped like horns contract when touched or when the object to which they cling is moved. Tufted polyps, on the other hand, are incapable of contracting. Touching or moving them, however, does vastly change their posture and their position. In fact they vanish quite suddenly, withdrawing entirely into a case (Plate 10, Fig. 8, i k l b i m*). This case is composed of a material similar to that making up the body parts I have already described and of which the body of the polyp is a product.

When the polyp has withdrawn into the case, its body (Fig. 8, i g b*) can be seen very distinctly through the transparent walls. In order to understand the position of a polyp when it has withdrawn into the case, one needs to know that the skin of the polyp is attached to the opening (Fig. 8, i b, i b) of the case in such a way that when the polyp moves inside, that portion of the skin cannot follow. Thus the skin remains attached by its lower end to the opening of the case which the polyp enters by turning itself inside out. The tuft, which holds by its base to the upper end (Fig. 8, a) of this skin, enters with it and ends up lodged in the tube (Fig. 8, a b) formed by the skin when the skin has entered completely and is completely inverted. The viscera sink down deeper into the case (Fig. 8, a g*) than any other parts do. Since both the opening in the case and the tube formed by the skin are much narrower than the tuft, the tuft must close in order to be able to enter. The arms come together as would the barbs of a feather when forced to enter a narrow tube (such as Fig. 8, a b*). If the polyp is left undisturbed, it will soon be seen emerging from the case where it has hidden. The arms appear first, initially clustered into a sheaf; but when they are about half-way out, they begin to draw apart at the tips. Finally the tuft opens, looking again as it did beforehand, and the body appears outside the case.

When the tufted polyps were well outside their case, I clearly saw a thread attached to the lower extremity of the stomach by one end and to the bottom of the case by the other (Plate 10, Fig. 8, g o). I saw another thread which seemed to be attached near the base of the tuft by one end, and also to the bottom of the case by the other end a o. Evidently these threads serve to draw the polyp into the case.

Rarely does one find a solitary tufted polyp. Ordinarily, there are a

*See Errata, Book I, p. 60.

number of them together, and in the species I am discussing they are arranged alongside one another. Several polyps often protrude from the same case (Fig. 8, *i k l* b *m*), but through separate openings (Fig. 8, *i b,* i b, *x y,* and Fig. 9). What I am going to say about their mode of multiplication will at the same time explain why they are found next to each other in this way.

In order to see the young unmistakably as they begin to grow, it is necessary to have quite a clear idea of the structure of the tufted polyps and to be experienced beforehand in observing them. First, a small elevation develops on the surface of the case of a mature polyp. Next one sees the young polyp's body and its tuft (Fig. 9, *r, r*) or rather, the base of its tuft (Fig. 8, *t s*) which is beginning to grow, and the tips of its arms (Fig. 8, *u, u, u*) which are emerging from the edges of this base. These arms grow in proportion with the growth of the body. The young polyp is ordinarily able to eat at the end of a few days. Its viscera, at first completely transparent, become brown after it has consumed some food.

When the water surrounding tufted polyps abounds with food, the young ones grow in great numbers. I have often seen more than a hundred joined together forming a very pretty bouquet. They then separate, but not one by one. The bouquet divides into two or three parts containing varying numbers of polyps (Plate 10, Fig. 9). This separation occurs quite imperceptibly. At first the mass formed by all the cases or, more correctly, by the common case, divides into two or three branches, and then these branches gradually separate from each other completely.

To observe conveniently what I have just described, I arranged it so that the bouquets of tufted polyps were fastened against the walls of a powder jar where I could observe them through a strong magnifying glass. In this way I not only saw them multiply and the various branches of their bouquets separate, but I also observed that these branches then move away from each other. Their progress is so slow that it is absolutely imperceptible. I have never seen a populous colony of polyps journey more than half an inch in a week. I have also observed a number of them that remained in the same spot for a very long time.

Previously I mentioned that the body of the polyps was an outgrowth of the case into which they withdraw. I wished to avoid giving the impression that the cases were constructed by the polyps in the way that caddis worms construct their cases. The cases should be considered a part of the polyps' body: they grow with it and in a similar fashion and are composed of the same materials, at least in the instances of the tufted polyps that I have studied.

What I have just said about the way in which young tufted polyps grow on the sides of mature polyps is enough to indicate the great similarity they bear in this respect to the polyps with arms shaped like horns; that is, they too multiply by shoots. Consequently, the tufted polyps furnish us a second example of this mode of reproduction among the animals.

In the preface to the sixth volume of his *Mémoires pour servir à l'Histoire des Insectes* (p. 76 and following), Mr. Réaumur furnishes yet additional examples of this mode of animal reproduction, namely, in the various species of tufted marine polyps. He has even discovered new ones since writing the preface, and it is more than likely that a large number of living organisms which have long been considered to be plants will likewise be found to be colonies full of polyps. This conjecture is rendered highly probable by the similarities shared by all these bodies that have been classified as marine plants, a number of which are assuredly masses of polyps.

I shall not enter into any greater detail on this matter. I know no better course to follow than to refer the reader to the remarks of Mr. Réaumur in the preface I have cited, and to any further information he and Mr. Bernard de Jussieu will present to the public on this interesting subject.

I believe I should mention one more fact concerning the tufted polyps of fresh water. Not only do they multiply by shoots, but they also produce eggs. We owe this information to Mr. Réaumur (the preface cited above, p. 76) who, together with Mr. Bernard de Jussieu, observed that the tufted freshwater polyps laid brown, slightly flattened eggs, from which these gentlemen saw young polyps born. I noticed small, white, transparent spherical bodies of various sizes on a number of the tufted polyps I was observing. I only suspected that these small objects were eggs, but I had no opportunity to investigate whether or not this conjecture had any foundation. These small objects were easily discernible through the transparent skin and case of the tufted polyp. They were in continual motion as though tossed from one place to another. I saw them pass from the case (Plate 10, Fig. 8, *i k l* b *m*) into the body of a polyp *a b i e,* move up between the skin and the viscera almost to the root of the tuft *e,* and then return from there into the case. What is more, those that emerged from the body of one polyp and passed into the case were not always pushed back again into the body of the same polyp, but rather into the bodies of various other polyps successively. I paid great heed to this phenomenon because it clearly proves that the cases of different polyps are interconnected, or in other words, that a number of these animals share a common case. If those spherical objects which I saw passing successively into the bodies of various polyps are eggs, it could be said that these eggs belong in common to all the polyps because their bodies are interconnected through their common case.

We saw previously (p. 118 and following) that the polyps with arms shaped like horns occasionally divide on their own, multiplying by sectioning themselves. As I have already mentioned, however, it seemed to me that this process occurred too infrequently for one to speak of it as an ordinary and natural mode of reproduction among the polyps, especially when compared to their method of multiplying by shoots.

There are animals, however, which, like the polyps, can multiply as the result of having been sectioned and which also can multiply by dividing on their own and do so quite frequently. One species of worms which I have already studied attentively propagates prolifically, and thus far I have seen them reproduce only by division. The animals in question are the barbed millepedes which I discussed in the second *Memoir* (Plate 6, Fig. 1; p. 52 and following). In the preface to the sixth volume of his *Memoirs* on the Insects (p. 59), Mr. Réaumur informs us of the results of his experiment in which he sectioned some of these millepedes. He saw each portion become a complete millepede. While performing the same experiment on these animals I learned that reproduction by division occurred in them not only when they were cut, but also naturally.

In May, 1741, I tried for the first time to cut a barbed millepede [*Stylaria;* see Book I, p. 48] into two pieces in order to see if it possessed the same property as did the polyps with arms shaped like horns that I was studying. I placed the two halves into a shallow glass vessel. They were easy to distinguish from one another because the first section terminated in a head quite conspicuous for its fleshy barb (Plate 6, Fig. 1, *d*) and for a black dot present on each side. Perhaps these two black dots are the eyes of the millepede. About a half-hour after separating these two sections, I returned to examine them with a magnifying glass. How great was my surprise when I saw that each section had a fully formed head! It seemed highly improbable that a head could have grown on the second portion of the millepede in so short a time; indeed, I could not convince myself that this was the explanation of the enigma puzzling me. Immediately, I set about cutting another millepede in half in order to try to assess the peculiar phenomenon that I had just seen. The same thing happened again. Shortly after the sectioning, I found that the two portions of this creature each had a head and that each was a complete millepede. In the same shallow glass in which I had placed the portions of the second millepede, however, I found in addition to the two sections which had heads, a third portion that was very short and had no head. Here was yet another riddle for me: to learn whence this third portion had originated, for I was absolutely certain that I had cut the millepede into only two pieces. I decided to observe whole millepedes more attentively than I had done previously, and by this means I found what I was seeking.

I saw a number of these animals which appeared to be not a single millepede, but two situated end to end. The head of the rear millepede, with its barb projecting upwards perpendicularly, was partially inserted into the last segment of the one in front. I concluded then that the millepedes I had sectioned were similar to these, and that the operation I had performed upon them had forced the two joined millepedes to separate. It seemed probable to me that the very short headless portion I had found in the glass containing the portion of the second millepede which had been cut in two, was nothing other than the end portion of the anterior millepede. Apparently I had cut this millepede a little distance from the place on its posterior extremity where the head of the rear millepede had been inserted. That is to say, an end portion of the anterior millepede had remained attached to the rear millepede; it soon thereafter had detached and had formed the third portion I had found in the glass.

Having discovered the facts just reported, I was curious to learn why a number of millepedes were found set end to end in twos. I systematically studied a number of them that were in this condition, and the variations which I noticed among them led me to suspect that these joined millepedes had never previously been separate. After observing these animals for some time, this is what I finally learned.

A single millepede may become a double one in a few days. About two-thirds down the body length, as measured from the anterior end, a head forms. The fleshy barb of the head is seen distinctly, growing perpendicularly upwards above the body of the millepede, the black dots appear on both sides of the head, and finally the new millepede, which is simply the posterior portion of the original millepede, is ready to separate and does separate on its own.

I was not content to observe the formation of some of these animals a few times in succession. I repeated the same set of observations frequently and always saw the same phenomenon. I removed some young millepedes from the vessel containing their mother at the very moment they separated and kept them in isolation as I raised them. I saw them grow, saw their posterior ends become millepedes, and saw them separate. I saw a single millepede produce a number of young, and I saw a number of these divisions occur successively within a few weeks. After a time the containers which at first held but a few millepedes were teeming with many.

From what I have just said about these animals, one can appreciate that they strongly warrant a careful study, as indeed I intend to initiate. Our present knowledge about them is sufficient to prove that they truly furnish an example of animals which multiply naturally by dividing themselves. Mr. Lyonet, the first person to cut worms and see each portion become a complete animal, also noticed that the worms he had cut could divide on their own. I believe I can make the same statement about those worms on which I fed the polyps for a long time. Because I have not made it my concern to study those worms attentively, however, I shall not attempt to describe the manner in which these animals multiply by dividing themselves.

PLATE 9

THE FIGURE which takes up this whole plate shows part of a piece of wood covered with long-armed polyps. I found it among a number of other such species in one of the ditches at Sorgvliet in July, 1742.

EXPLANATION OF
THE FIGURES OF THE THIRD MEMOIR
P L A T E   8

FIGURE 1 depicts a polyp of the second species from which two young polyps are emerging. The one at *e* is as yet no more than a simple conical-shaped bud; the other *i c* is nearly cylindrical with arms already showing at its anterior end *c*.

In Figure 2, a polyp of the second species is shown bearing a young one *a b* which is ready to separate.

Figures 3 and 4 illustrate the manner by which young polyps separate from their mother. In Figure 3, *a b*, a mother is shown fastened to some object by her extremities *a* and *b*. The body of the mother forms almost an arc *a d b*. The offspring *c d*, attached to its mother at *d*, is fastened to the glass or some other object by its anterior end *c*. In this situation the mother can contract its body so that it now becomes the chord *a b*, Figure 4, from that circular arc *a d b*, Figure 3, that it fashioned earlier. The young polyp, remaining attached at *c*, Figure 4, does not follow the mother and is compelled to separate. The two turn out to be at some distance from each other immediately upon separating.

Figure 5 shows the inner surface of a piece of skin taken from a polyp on which a young polyp is growing. The connecting opening between the mother and the offspring is seen at *t*, and *o* is the opening at the end of the remaining portion of the offspring after its anterior half has been removed.

Figure 6 depicts the same specimen as Figure 5, but as it appears on the outside. The remaining portion of the offspring is at *e; i* indicates the opening communicating between the mother and her young.

In Figure 7 a polyp *a b* of the second species is seen as it appears some time after it has eaten a worm. Part of the food substance has passed into the young ones *d, c, e, i, o* and has made them swell a little.

Figure 8 portrays a polyp *a c b* of the third species, two weeks old and already bearing nineteen little ones, including a third generation produced by the offspring of the polyp.

Figure 9 shows the bud *e* of Figure 1 severed from the mother polyp.

Figure 10 is the same bud as seen under a microscope.

Figure 11 presents a polyp *a b* of the third species, remarkable for one of its arms, *a f f f f f*, which is branched in a number of places, namely at *c, d, e,* and *i*.

Pl. 8. Mem. 3.

Fig. 1.

Fig. 2.

Fig. 3.

Fig. 8.

Fig. 4.

Fig. 11.

Fig. 7.

Fig. 5.

Fig. 6.

Fig. 9.

Fig. 10.

P. Lyonet delin. et sculp. 1743.

PLATE   10

IN FIGURE 1 an aquatic caddis worm [see Book I, p. 48] is shown swimming with eight long-armed polyps attached to its case by their posterior ends. I have seen a number of polyps which were not induced to contract by the motion of the caddis worm any more than those shown in this figure.

Figure 2 depicts a polyp of the second species on which there are two little spherical bodies *e, e.*

In Figure 3 we see a polyp of the second species about an inch and a half long. It has eighteen arms around its mouth and, in addition, one other *c d,* which is displaced.

Figure 4 shows a polyp of the third species, its portion *a c* covered with nearly pyramidal white excrescences that have emerged from its body.

Figure 5 illustrates a polyp *a c* with two heads or, if one prefers, two polyps *a m, c m* situated end to end and joined together. This polyp has developed from an offspring that sprouted two heads simultaneously at its anterior end.

Figure 6 shows a polyp of the second species bearing three conical young *c, c, c,* each of which has but a single arm at the tip of the cone.

I have kept the polyp shown in Figure 7 for more than two years. About a year and a half after its birth, it grew seven extraordinary tails *q, q, q, q, q, q, q* and three conical young *c, c, c.*

Figures 8 and 9 illustrate tufted freshwater polyps [*Lophopus;* see Book I, p. 48]. Three of them enlarged under the microscope are portrayed in Figure 8: namely, one *b f a c d d d e h g i,* which is outside the case *i k g m b;* another i b a g, which is inside the case *i m g l b* and the young offspring *t x y s,* which is also on the outside. The polyp's body *a b i e* is visible through the transparent skin, and one can distinguish the esophagus *e h,* the stomach *f g,* and the rectum *a f.* The polyp's skin is attached to the opening *i b* of the case. The polyp's tuft, or head *a c d d d e,* is composed of the base *e a c* of the tuft which is barely visible, and the arms *a d, a d, a d,* which emerge from the edges of this base. The letters i b a g indicate a polyp which has withdrawn into its case; i a b depicts the skin of the polyp turned inside out and in which the tuft is lodged. The viscera are between a and g. As for the young polyp, *t s* is the base of its tuft, and *u, u, u* are the arms which are beginning to sprout. The threads *g o, f o* are attached at one end to the viscera, and at the other to the bottom of the case. In this figure they are depicted as being detached or cut; for this reason they are not stretched taut to *o, o.* Each polyp has its own opening, *i b,* i b, and *x y,* into its own case.

Figure 9 pictures a bouquet of tufted polyps, life-size, fastened to a piece of wood *a b* by the base *c d* of the polyp colony. This base *c d* is but a mass of material which served as a case for the polyps but which lost that function when the colony increased in size. Frequently one finds polyp colonies that have no such base at all. Shown life-size are the tufts *p, p, p,* the polyp's body *i e,* and young polyps *r, r.* In addition, this figure shows the polyp colony as having begun to separate into three branches, one of which is ready to detach completely from the other two.

Pl. 9. Mem. 3.

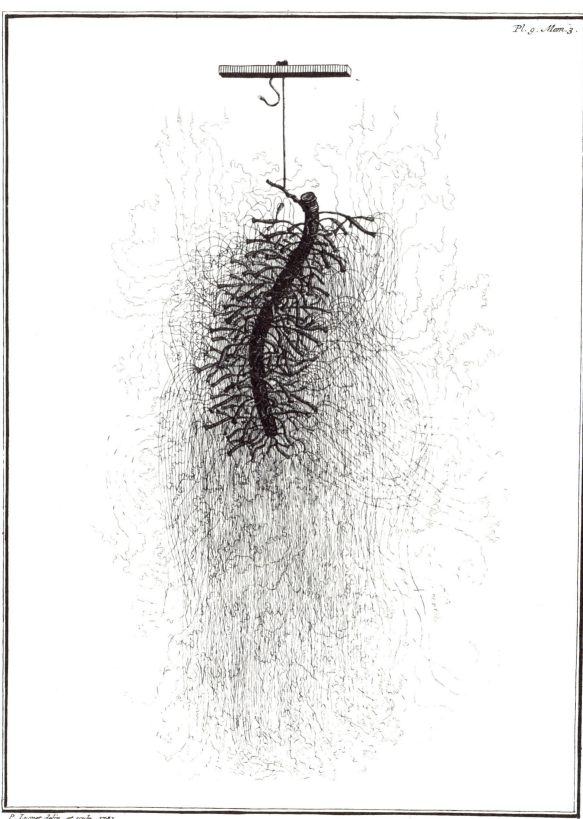

MÉMOIRES
POUR L'HISTOIRE
DES POLYPES.

✱✱✱✱✱✱✱✱✱✱✱✱✱✱✱✱✱✱✱✱✱✱✱✱✱✱✱✱✱✱✱✱✱✱✱✱✱

QUATRIÉME MÉMOIRE.

Opérations faites fur les Polypes, & les fuccès qu'elles ont eu.

A première opération, que j'ai faite fur les Polypes, a été de les couper *transverfalement.* On a vu en general au commencement du premier Mémoire ✱, quel en a ✱ Pag. été le fuccès. J'ai renvoié à celui-ci le détail de cet- &c. te Expérience.

POUR

Pl. 10. Mem. 3.

Fig. 3.

Fig. 2.

Fig. 1.

Fig. 5.

Fig. 4.

Fig. 6.

Fig. 7.

Fig. 9.

Fig. 8.

MEMOIRS

CONCERNING THE NATURAL HISTORY

OF THE POLYPS.

✳ ✳ ✳ ✳ ✳ ✳ ✳ ✳ ✳ ✳ ✳

FOURTH MEMOIR.

*Operations performed on the polyps, and
the results obtained.*

he first operation I performed on the polyps was to cut them *transversely*. The general results of this experiment were presented at the beginning of the first *Memoir* (p. 8 and following). I have postponed the details until now.

To cut a polyp transversely, I place it with a little water in the hollow of my left hand. At first it lies contracted at the bottom of the water. Even when it is in this state it can be cut through easily enough provided one uses a very fine scissors. When the polyp is elongated, however, it is easier to divide it precisely where one wishes. I therefore usually keep the hand containing the specimen still for a moment to allow the polyp time to elongate. When I have it as I want it, I delicately pass one blade of the scissors, which I hold in my right hand, under the part of the polyp's body where it is to be severed. Then I close the scissors, and immediately after having divided the polyp, I examine the two halves under the magnifying glass in order to assess the outcome of the operation. If it has succeeded, I put each of the two portions of the polyp into separate vessels or both together into the same one. One may place the two halves in the same container without fear of confusing them until the posterior part has nearly completed its regeneration.

I placed the parts of the polyps on which I performed my experiments into shallow glass vessels containing only nine to eleven millimeters of water. By this method I could always observe these pieces with a magnifying glass in whatever part of the container they might be.

Since I had a number of vessels at the same time in which I was keeping severed polyps, I marked each one with a number or a letter, using the same symbol in the journal of my observations to identify these polyps. I was the only person to handle the vessels and, when putting in fresh water, I was extremely careful not to mix anything. I took these same precautions with all the polyps upon which I performed the experiments to be reported in this *Memoir*.

The two halves of a polyp, just after being separated and placed in a glass, at first lie contracted at the bottom of the vessel. Ordinarily they do not remain there for long without elongating to a lesser or greater extent. The head of the first part (Plate 11, Fig. 1, *a*) is the original head of the polyp that was divided. Even shortly after the sectioning, this part scarcely differs from a complete creature, except that its posterior extremity (Fig. 1, *c*) is a little wider than that of a normal polyp and it has a very obvious opening. As this first part elongates, the opening at its posterior end closes and the posterior extremity narrows and becomes like that of a fully formed polyp. During summer it often happens that the first part walks and eats on the very day on which it has been separated from the other. I have even seen some which fed immediately after the sectioning.

The second, or hind part (Plate 11, Fig. 2), after elongating a little, usually is open at its anterior extremity *c,* with the edges of the opening turned slightly outward. These edges next fold inward and the fold which they form serves to close the aforementioned opening. The anterior end then appears simply swollen (Fig. 3, *c*) and usually remains more or less in the same condition until the regeneration destined to take place there is accomplished.

I have never seen such a hind part change its position before the conclusion of this regeneration. Most of those that I have observed remained lying on the bottom of the glass in the same place, in various stages of contraction or elongation, and with their posterior extremities attached to the glass. A few stood perpendicularly erect on the bottom of the glass.

The arms which sprout at the anterior end of the hind part grow precisely as do those of young polyps. First, one sees the tips of three or four emerging from the edges of this extremity (Plate 11, Fig. 4, *c*). While these first ones are growing, others appear in the spaces left between them. Even before these arms have completed their growth they can seize prey; and since by that time the mouth is fully formed, the prey is swallowed as it would be by a complete polyp.

The regeneration I have just described takes place more or less quickly according to the degree of warmth. At the height of summer I have seen severed hind parts, the arms of which began to grow after twenty-four hours, and which in two days were capable of eating. In cold weather, on the other hand, I have seen some that required fifteen to twenty days for the heads to form.

It seemed to me that, all other things being equal, regeneration would occur more rapidly in the hind part of a well-nourished polyp than of a polyp which has been without food for some time.

As soon as a second or hind part has acquired a head, that is, as soon as the mouth is formed and the arms have grown, it looks, as does the regenerated first part, exactly like a polyp which has never been divided; thus both halves can be considered complete polyps. They exhibit all the known properties of these animals: they perform all the movements of which polyps are capable; they seize prey; they feed on it; they grow and multiply.

If a transverse cut is made on polyps which are developing young ones, those young will continue to grow after the section is made on whatever part of the sectioned polyp they turn out to be. It happens often that hind parts which had no young at all at the time of sectioning develop some young even before they acquire arms and before they have been able to eat. It seemed to me that the development of the offspring sometimes retards that of the regenerating arms. I observed one young polyp emerge so close to the anterior end of a severed second part that, after it had grown a little, it merged with the end on which no head had yet developed. This young polyp appeared to be the anterior end of the severed hind part, but it bent a little and made a markedly obtuse angle with the hind portion instead of a straight line. When the head of the young polyp was able to eat, I gave it some worms which passed from its stomach into the portion of the polyp from which it had sprung just as prey passes from the mouth of a whole polyp into its stomach. Some time afterward, the posterior extremity of the young polyp formed a constriction and then broke away. For several days longer the severed hind part remained without arms; finally arms began to appear at its anterior end, and six of them developed.

Not satisfied with knowing that sections of polyps that had been cut into two would walk, eat, and multiply, I still wished to ascertain whether they had the same propensity for light I had observed in polyps that had not been sectioned (see *Memoir* I, pp. 7 and 37). Therefore, I cut in two a con-siderable number of polyps, placing the first parts into one glass and the hind ones into another. From experiments repeated very frequently, I saw that both the first and the hind parts sought the best-lit area of the jar.

In whatever area of the body a polyp is sectioned, whether at the middle or more or less close to the anterior end or to the posterior end, the experiment succeeds equally well; the parts uniformly become complete polyps. This holds true even in the case of the polyps with long arms when they are cut below the point where their bodies begin to narrow; that is, when one separates a part of this tail from the rest of the body, even it becomes a complete polyp. Actually, some time is required for this to occur and occasionally it happens that the tail part of a polyp accidently dies before regeneration can take place.

Frequently I have cut merely the edges off the anterior end of a polyp, that is, the circle from which the arms emerge (Plate 11, Fig. 5). No matter how narrow it was, a polyp developed from it. At the beginning, this polyp was all arms, but later when I took care to feed it, the body became as large as that of other polyps. Sometimes such polyps assumed a somewhat irregular shape at first. Some of their arms were located elsewhere than at their anterior end, as already mentioned in the preceding *Memoir*.

What is more, even parts of such circles as those just discussed, pieces which had only two (Plate 11, Fig. 6) or three arms, became fully formed polyps. New arms developed to make up for the missing ones.

I severed some arms and observed them to see whether they would ever develop into polyps. The experiment did not succeed. I would not wish to conclude, however, that successful regeneration from a single arm is impossible.

The preceding statements will already have indicated that a polyp can be divided successfully into more than two parts. I have cut some transversely at the same moment into three and four parts, and all these parts have become complete polyps.

The first and the last parts of polyps cut transversely into three or four pieces are just like the first and second halves of a polyp which has been cut in two. The posterior end of the first part must form itself into a tail, whereas a head must develop at the anterior end of the last part. A double regeneration must occur in the intermediate parts, however; that is, in the second part of a polyp cut into thirds, and in the second and third parts of one cut into quarters. I repeat: the regeneration that must take place in these portions of polyps is double, for these are stumps which have neither tail nor head and must acquire both. Indeed, this is what does happen within a length of time that varies with the circumstances. In order to convey a more precise idea of these experiments, I will present excerpts from various entries in my journal concerning polyps sectioned into three and four parts.

I sectioned a polyp transversely into thirds on July 18, 1741. The first part did well. It had the six original arms of the polyp I had sectioned.

The second part began to sprout an offspring on the 23rd; as yet it had no arms at all. On the 25th I noticed arms both on the young polyp and on the second part as well. The young polyp separated on the 28th. At that time the middle part had seven arms. Arms began to grow on the third part on the 24th. I perceived five on the 25th and seven on the 28th.

These three parts of the same polyp ate water fleas on July 29th. They were complete polyps.

I divided another polyp transversely into three parts on July 15th, 1741.

The first part had seven arms.

Arms began to appear on the second part on the 25th. On the 28th it had eight of them.

On the 22nd they began to grow on the third part. I saw seven of them on the 25th and eight on the 28th.

These three parts from the same polyp ate water fleas on July 29th.

On July 18th, 1741, I sectioned a polyp transversely into four parts.

The first part did well. It had seven arms.

Arms started to grow on the second part on the 25th. I noticed five of them on that date, six on the 28th, and seven on the 29th.

Arms began to appear on the third part on July 25th. I noted five on that day, six on the 28th, and eight on the 29th.

The arms of the fourth part began to grow on the 23rd, two days earlier than those of the two preceding parts. On the 25th I saw three of them that were already quite long. It had six on the 29th.

The polyps are not large enough to be divided into a very great number of parts all at once. I compensated for this by successively sectioning a polyp into many parts. After having cut one polyp into four, I took care to feed each of these four parts. When they reached a certain size, I cut them again into two or three parts as their dimensions allowed. Next I likewise grew all these parts and sectioned them again. In this manner I cut the polyp in question into fifty parts. There I stopped, believing that it was sufficient to have pursued the sectioning to this point. All fifty parts became fully-formed polyps. I saw them perform all their functions. I kept many of them for more than two years, and they multiplied greatly. I performed this experiment upon a polyp of the second species.

For more than two years without interruption, I raised polyps that had never been sectioned, and also polyps which had developed from parts of sectioned polyps. All circumstances being otherwise equal, I did not find that one group multiplied more than the other.

A number of times I tried sectioning young polyps transversely while they were still attached to their mother, an experiment which succeeded very well. I shall mention only a few particular cases.

I sectioned transversely three offspring which were emerging from a polyp of the second species, cutting off the mother's head at the same time. In a few days, I am not able to say precisely how many, a new head grew on the mother and on the hind parts of the three young which were still attached to her body. Subsequently these young polyps separated from the mother and became complete polyps. The four heads, that is those of the mother and of the three offspring, which had been severed from the rest of their bodies, also developed into complete animals.

On May 26th, 1741, I cut off the head of a polyp of the second species as well as the heads of two offspring that were emerging from its body. On the 30th, arms began to appear on the hind part of the mother and on the hind parts of the two young which were still attached to her body. The three severed heads also grew into complete polyps.

On June 23rd, 1741, I cut off the head of a polyp with long arms and also the heads of two offspring it was bearing. On the 25th, arms appeared on the rear part of the mother and on that of the young still attached to her body. The young separated from the mother during the night of June 26th to 27th. They were then already complete animals. The heads which had been cut from the mother and the young also became complete polyps.

After having cut a great number of polyps transversely, I undertook to cut them *lengthwise,* that is to make the cut parallel to the length of their bodies. This operation is a little more difficult than sectioning them transversely, but it can be accomplished readily by taking a few precautions.

One has much less of a grip on a polyp that one is cutting lengthwise when it is elongated than on one being cut transversely.

To gain a hold on a polyp which is to be cut lengthwise is a matter of making it widen. To bring this about it is necessary to make it contract as much as possible because the more contracted the body of a polyp, the wider it is. I place a little water and the polyp that I wish to cut lengthwise into the hollow of my left hand. Instead of leaving it in this water and simply making it contract there, however, I guide it with the tip of a brush to the edge of the water. It then is resting on my hand, where I try to make it contract as much as possible. It is to be noted that a polyp widens more when it contracts outside the water than in the water, because outside the water it flattens itself somewhat against the object upon which it rests.

This is not the only reason that it is preferable for the polyp which is to be cut lengthwise to be out of water rather than in it. When it is in the water the least motion one makes, whether by a movement of the hand holding the polyp or by passing the scissors under its body in order to cut it, causes the polyp to change its position, with the frequent result that one does not cut in the location desired. On the other hand, when a polyp is out of the water one can easily pass the scissors under its body without disturbing it.

There is a further expedient one can add to those just indicated for widening the polyps which one wishes to cut lengthwise: that is to feed them well before undertaking the operation. As shown in the second *Memoir,* food forces the body of a polyp to widen greatly. The polyp can be placed on the hand when its stomach is full and can be cut with the food in its body. It is not enough to have the polyp thoroughly contracted and widened on the hand. In addition, one must take great care to arrange the polyp in such a way that it forms a straight line and that its two extremities are easy to distinguish.

Having made all these preparations, I take a pair of very fine scissors in my right hand. Beginning with the point of one of the blades, I pass it under the polyp, working it so that this blade lies under the polyp parallel to the length of its body. One can begin to pass the blade of the scissors under the body of the polyp at whichever of the two ends of the body one wishes. But since the anterior end is usually wider, it is more advantageous to start there, the more so because one can then more easily section the head evenly. When the blade of the scissors which I pass under the polyp is parallel to it lengthwise, and when I see that its point extends ever so slightly beyond the end of the polyp opposite from where I began to pass the blade under the body, I close the scissors. Ordinarily the polyp turns out to be sectioned into two quite equal longitudinal parts.

As I discussed in the first *Memoir* (p. 28), a complete polyp forms a sort of tube extending from one end of the body to the other. A half of a polyp which has been cut in two lengthwise (Plate 11, Fig. 7) therefore forms half of a tube, its anterior extremity (Fig. 7, *a e*) terminating in half of the head, that is, half of the mouth and half or nearly half the number of arms.

After having cut the polyp, I tilt my hand very slightly so that the water in the hollow of the palm is brought over to moisten the body. Then, with the dampened tip of a brush, I separate the pieces of the polyp from each other, spreading them out on my moistened hand in order to examine them conveniently through the magnifying glass. Each part then appears to be a small strip of the polyp's skin, terminating at one end in several contracted arms. I take the precaution of moistening my hand, the brush, and the sections of the newly cut polyp so as not to risk tearing these sections, as can easily happen without this precaution. After having examined the cut parts of polyp, I guide them with the brush into the water in the hollow of my hand. From there, I lift them out one after the other and place them in a flat glass vessel in order to continue observing them.

When a polyp has been cut lengthwise into two, most of the time the halves twist themselves about at first and curl up in various ways. Usually they begin to roll up at one end, forming a very tight coil (Plate 11, Fig. 8). I have noticed that the external surface of the polyp's skin is always on the inside of the coil. Then the coil straightens out; and, after a longer or shorter interval, the piece of polyp forms a tube as the edges of the two sides (Fig. 7, *a i, e b*) draw near each other and join. Sometimes it happens that the edges begin to join together first at the posterior end; at other times the edges draw together nearly everywhere all at once. When the joining is gradual and begins at one end, the closed portion of the polyp (Fig. 9, *c i b*) can usually be distinguished from the one not yet closed (Fig. 9, *c a e*). Frequently, the section of polyp is so twisted when the edges are beginning to join that one cannot see how they draw together.

In reuniting, the edges join so well that from the very first moment no scar can be seen in the area where they have joined. The skin of this new polyp is as smooth there as it is elsewhere. Consequently, as soon as the joining of the edges is completed, the bodies of these parts of polyps resemble those of complete polyps (Fig. 10, *a b*). The heads bear fewer arms, however, each having only three or four depending on the number of arms possessed by the polyp which was sectioned and depending on how the section was made.

Usually the parts of a polyp sectioned lengthwise require only an hour, sometimes even less, to take on the form of a complete polyp.

Although the parts from polyps sectioned lengthwise have only a few arms, they suffice for them to seize prey. I was at first surprised to see some of them seize and swallow a worm as long as themselves 24 hours after they had been sectioned. Subsequently, I attempted to feed some halves of longitudinally sectioned polyps even sooner, and observed some that ate three hours after sectioning. I have reason to suspect that parts from a polyp which has not had food for a certain time prior to sectioning eat sooner after the operation than those from a well nourished polyp.

I have just stated that polyps developing out of the sections cut longitudinally have only a few arms at first. In a few days, however, others begin to appear in the places where they are lacking (Fig. 10). When these attain the length of the original arms, there is absolutely no difference between these polyps and those which have not been cut.

I have also sectioned polyps into four longitudinal parts at the same time. Here is how I went about it: after having cut the polyp in two and having spread the two halves out on my hand, I cut these two halves in the same way that I cut the whole polyp. In a short time, each of these four sections of a polyp severed lengthwise developed into a complete polyp.

The following excerpt from my journal contains an account of the experiment in question. On the 16th of August, 1741, I sectioned a polyp of the second species, which at the time had eight arms, into four longitudinal parts. I called these four parts numbers 1, 2, 3, and 4.

Immediately after the sectioning, No. 1 had three arms. A new one appeared on the 18th, another on the 20th, one on the 23rd, and one more afterward.

No. 2 had two arms. Three new ones appeared on the 18th, one on the 22nd, and one more afterward.

No. 3 had two arms. Three new ones appeared on the 18th , one on the 20th, and one more afterward.

No. 4 had one arm. Three new ones appeared on the 18th, one on the 20th, one on the 22nd, and one more afterward.

Six days after the sectioning the four parts of the one polyp had six arms each, and by September 14th each had seven. All the parts ate and multiplied. Subsequently I cut them transversely, and the experiment succeeded perfectly.

When one cuts polyps lengthwise that are producing young, these young continue to grow after the sectioning.

In one experiment, I performed upon a single polyp both of the operations under consideration until now; that is, I cut it at the same time both transversely and lengthwise, in a fashion quartering it. The four pieces of this polyp became complete polyps. For this to take place, each section had to undergo both types of regeneration: the one which I described as occurring in polyps cut transversely and the one occurring in those cut longitudinally.

The two first parts of the quartered polyp, those pieces furnished with some arms at their anterior extremity, first formed into tubes as their edges drew together and joined; secondly the posterior extremities of these first parts assumed the shape of a polyp that had never been cut. As for the two hind parts, once the edges had drawn together to form a tube, they were in the same condition as the hind part of a polyp which had been cut transversely a short while before. Their development then proceeded in the same manner as did regeneration in the hind part sectioned from a complete polyp.

When the four polyps that grew out of the quartered sections were able to eat, I gave them food. Then I quartered one of them, making four polyps out of it. After a certain time, I again performed the same operation on one of these four polyps. I maintained these last four polyps for a period of two years and saw them multiply.

I cut one polyp lengthwise part way, starting at the head, and it eventually came to have two bodies, two heads, and one tail. Each portion of the polyp which had been separated, instead of drawing together to form a single polyp, formed a separate head and a body of its own, as I have said the parts of a polyp do when sectioned lengthwise from one end to the other. I fed this two-headed polyp, making it take food through both mouths. Afterwards, starting at the head ends, I also split each of its two branches part way. In a short time it had four heads. Finally I succeeded in making it acquire seven heads (Plate 11, Fig. 11). Another polyp on which I performed the same experiment acquired eight heads.

I have seen these Hydras taking food through all their mouths at the same time.

The reader may well imagine that after I had succeeded in making some Hydras, I was not content to stop at that. I cut off the heads of the one that had seven, and a few days later I beheld in it a prodigy hardly inferior to the fabulous Hydra of Lerna. It acquired seven new heads; and, if I had continued to cut them off as they sprouted, no doubt I would have seen others grow. But here is something more than the fable dared invent: the seven heads that I cut from this Hydra, after being fed, became perfect animals; if I so chose, I could turn each of them into a Hydra.

If one begins to cut a polyp lengthwise at the posterior end and cuts it only up to a point near the head, a short time later the polyp will turn out to have one head and two tails. The number of these tails can be augmented by cutting them again after they have been made to grow, as I have said I did with regard to the Hydra with many heads. I have seen both the Hydras with many heads and those with many tails walk.

I have produced Hydras by a somewhat different method from the one just described. Instead of splitting the polyp lengthwise up to a point near one of its extremities, I open it from one end to the other and spread it out thus opened on my hand. In the first *Memoir* (p. 30) the procedure I use in performing this operation is presented, and Figure 7 of Plate 4 of that *Memoir* depicts, enlarged under a microscope, an opened polyp with its skin spread apart. With the scissors I snip the skin on all sides and in every direction, but taking care not to separate the pieces of skin completely and making certain that they are all still held together at some point. One can well imagine the condition of the polyp following all this. I have often seen those that I had cut up in this manner later develop three or four heads and sometimes some tails as well. Each of these little bits of skin, which are held to the others by only one of their edges, ordinarily becomes either a head or a tail. Possibly some of the pieces reunite, but I have not observed this.

Most of the time the heads and tails of the extraordinary polyps I have just described break away on their own and become complete polyps. I kept one seven-headed Hydra for six weeks before any of its heads separated. When the many-headed and the many-tailed polyps are well fed, they multiply like ordinary polyps.

The reader may already understand from the last experiment just reported, in which I slit a polyp open from end to end and snipped it repeatedly with the scissors, that no matter how nor in what direction one cuts a polyp, one simply does not kill it. On the contrary, one produces many from a single specimen. In order to be more certain of this fact, I pushed the operation just described still further. I slit a polyp open on my hand, spread it out, and cut in every direction the single piece it then formed, reducing it to little pieces, as it were, mincing it. These little bits of skin, both those that had arms and those that had none at all, developed into complete polyps. It does happen at times, nevertheless, that one sees some die. Perhaps death occurs because they are too small, or perhaps because of some accident which would have been equally fatal to a fully grown polyp.

These pieces of skin of which I speak should be observed carefully. It is necessary to examine them often, and when they begin to be able to eat they must be given extremely small portions of a worm. Above all, however, it is necessary to take precautions when changing their water; otherwise one risks losing them.

I have repeated this experiment successfully a number of times.

All polyps that develop from a piece of a polyp that has been cut in two or into many parts in whatever manner, always have a canal extending from their anterior to their posterior end; that is, they always have a stomach just as do polyps which have never been sectioned. For a long time I was at a loss to understand how such a stomach forms in the polyps which develop from a very small, single piece of skin which is narrow and short, or in those which grow out of a polyp sliced lengthwise into four or five parts, that is, in polyps which immediately after the sectioning are only bits of simple skin, thin and long in proportion to their thickness. It is easily seen that no matter how small the pieces of a polyp cut transversely may be, they always form a kind of tube, an extremely short one indeed, but one which elongates proportionally as these pieces of polyp lengthen. The parts of a polyp cut lengthwise into two remain wide enough for the edges of the opposite sides to draw together so that each of these pieces which was at first, so to speak, a half tube, then becomes a whole tube. I have seen this very often, as I have said previously.

But is the process the same, for example, with parts of a polyp cut lengthwise into a greater number of extremely narrow pieces? For a long time I remained in doubt about that question. One part of a polyp sectioned lengthwise appeared to me to have developed into a complete polyp without its edges drawing together and uniting. I had reason to believe that this piece of skin had simply swelled out and by this means had assumed the shape of a polyp. I saw this occur only once, however, and then not clearly enough to base a decision upon it. Consequently I set to work to repeat the same experiment.

I began by first observing extremely narrow strips of polyps cut lengthwise. I placed these sections into shallow glass containers where I could watch them through a strong magnifying glass. It is very important to examine them often, especially during the first two hours after the sectioning, in order to note carefully the different shapes they assume and to be able to ascertain that the edges of the opposite sides have not drawn together and united. I am going to describe, therefore, what I have seen many times in succession, inasmuch as I repeated the same experiment frequently for greater certainty.

An extremely narrow strip of a longitudinally sectioned polyp usually begins by coiling up in the same way as I have said the wider pieces do. It forms a very thin coil, the coiling always starting at one of its two extremities which then ends up in the center of the coil. As one can visualize, the coil is not composed of many turns, and indeed, often it is made up of only a single turn. The strip of polyp forms a circle, with its two extremities touching. Next the coil unwinds. I have seen some alternately roll up and then unwind two or three times during the first hours following the sectioning. There are pieces, on the other hand, which stretch out at once and remain extended. By observing such a piece of polyp often and always remembering the condition in which one left it when one broke off watching it, one may easily ascertain that: the piece definitely does not fashion itself into a trough from one end to the other, the sides do not draw close and join, and, in short, it definitely does not take the form of a tube. I have seen some that I am certain never took this shape. These strips of polyps inflated little by little. Narrow and flat at the start, they afterwards swelled until they became nearly cylindrical. Observing them in this condition through a magnifying glass in broad daylight or by candlelight, one can perceive already inside them a hollow space that extends from one end to the other. Afterward there is indeed reason to be further convinced of it. A head forms at one end of these parts of polyps, and in a few days they are in a condition to feed on little pieces of worm. One sees the pieces of worm as they enter the bodies of the polyps and can easily observe that they fill the hollow tube which has been formed inside. Finally the parts of polyps become complete animals.

The way in which the stomach of these parts of the polyp is formed is not the only remarkable phenomenon they present to us. The external surface of their skin is also quite worthy of attention. It is easy to understand that the internal surface of a piece of the polyp's skin is a part of the stomach wall of the polyp. Thus, when the piece of skin does not form a tube but simply swells out in order to become a polyp, that side which had been part of the internal surface of the stomach in the whole polyp before it was sectioned becomes a part of the external surface of the new polyp.

The following is an excerpt from a single item in my journal concerning the pieces of polyps just discussed. On November 1st, 1743, between one and two o'clock in the afternoon, I cut a polyp lengthwise into a number of parts. Here are the observations I made on one of them: This piece of skin was rather long and extremely narrow and was slightly coiled. Observing it without interruption, I saw clearly that the edges did not join together. By four-thirty it had already assumed a cylindrical shape, that is, it was already noticeably inflated. It had no arms at all. On November 3rd, there appeared at some distance from its most pointed end an outgrowth resembling the bud of a young polyp. On the 4th I noted that it was indeed a bud. Arms were even emerging from the anterior end of the offspring as well as from the mother, that is from the sectioned piece of polyp. Subsequently, the mother portion ate and multiplied.

I also made uninterrupted observations on extremely small pieces of skin which were so short and narrow that it was in no way possible for them to coil into a tubular shape. They swell out, become round or nearly round, and take on the shape of a bead that is hollow on the inside. Next they sprout arms at a spot on their surface in the middle of which is the mouth. Then they walk, they elongate, and in the end they are perfect polyps.

Now I come to an experiment no less curious than those in question thus far. This experiment consists of *turning polyps inside out*.

Here it must be recalled once more that the entire body of a polyp forms nothing but a tube, a sort of gut or pouch which extends from one of its extremities to the other. We are concerned then with turning this gut formed by the body of a polyp inside out as one would a pouch, a stocking, a glove, or the finger of a glove, so that the interior surface of the polyp becomes the exterior surface and the exterior surface becomes the interior.

Had I known that a piece of a polyp's skin could become a complete polyp merely by swelling so that a hollow tube forms in the center and becomes the animal's stomach, I would have had greater expectations of seeing an inverted polyp survive. I would already have had evidence that the internal surface of the polyp's skin can become, at least in part, the external surface, as described in the preceding experiments. When I undertook to turn polyps inside out, however, I had not yet performed those other experiments.

I thought of turning polyps inside out after noticing a phenomenon detailed in the second *Memoir* (p. 80 and following): namely, that the granules or vesicles which bedeck the entire skin of these animals become filled with nutrient juice. It occurred to me that if the granules which were on the external surface of the skin were closer to this nutrient juice, they would be the first to become filled with it, and that perhaps the polyp would be nourished as thoroughly as when the juices pass first into the vesicles lining the walls of the stomach.

The first idea that occurred to me was to place a polyp into a liquid solution which could be regarded as a nutrient juice for it. I wished to see whether the polyp possibly would absorb nourishment through all the external parts of its body, as do a number of marine plants according to Count Marsigli in his *Histoire de la Mer*. He reports that these are all full of small glands or vesicles which draw in the substance that nourishes them externally from the sea water bathing them. It is true that recent discoveries on these presumed marine plants alter this hypothesis a great deal, but they had not yet been made in July of 1741 when I was thinking about performing the experiment on the polyps I am now discussing.

Since I had been unsuccessful in finding a fluid in which I could place the polyps that would serve to nourish them, I thought of inverting them so that the external surface of their skin would form the walls of their stomachs. I had very little confidence that I would see this experiment succeed, but I did not believe it proper not to try it.

I attempted to turn polyps inside out for the first time in July, 1741. In the process I employed all the methods I could then invent, but to no avail. I was more fortunate the following year when I finally found a rather simple expedient.

So long as my attempts to invert them were made on polyps with empty stomachs, they came to nought. To the contrary, I had immediate success as soon as I fed them well before the operation so that their bodies were greatly widened. It is indeed prerequisite for a favorable result that the stomach and the mouth of the polyp be quite wide. For this purpose I first used the red larvae of the midge [*Chironomus;* see Book I, p. 48] mentioned in the second *Memoir* (p. 63). I also successfully used the same worms [*Tubifex;* see Book I, p. 48] I usually fed to the polyps in winter (Plate 7, Fig. 2). The first polyps that I turned inside out belonged to the second species.

I start then by giving a worm to the polyp on which I wish to perform this experiment; when the worm has been swallowed, I can begin. It is not advisable to wait until it is nearly digested. I place the polyp, its stomach quite full, in a little water in the hollow of my left hand. Then I press on it with a small brush nearer to the posterior than to the anterior extremity. In this manner I push the worm that is inside the stomach against the polyp's mouth which is forced to open. By pressing on the polyp a bit more with my brush, I make the worm (Plate 11, Fig. 12, *c e*) emerge partially through the mouth (Fig. 12, *a*) and to that extent I empty the back portion of the stomach. This worm emerging from the mouth of the polyp forces it to widen considerably, especially if it emerges bent double, as shown in the figure just cited.

When the polyp is in this condition, I guide it out of the water gently without disarranging it at all and place it on the edge of my hand, which is moistened only to prevent the polyp from clinging to it too much. I force the polyp to contract more and more, thereby helping to make its mouth and stomach widen even further. One should recall here that the worm, emerging in part from the mouth, is holding it open (Fig. 12).

Next in my right hand I take a bristle from a boar or a hog, one that is fairly thick and not pointed. I hold it as one does a lancet to let blood. I bring the larger end of the bristle (Fig. 12) to the posterior extremity of the polyp; I push on this extremity and I make the extremity enter the stomach of the polyp. This is accomplished all the more easily because the rear portion of the stomach is empty and stretched very wide. I continue to advance the end of the bristle which, as it moves forward, turns the polyp ever further inside out. When it reaches the worm which is holding the mouth open, the bristle either pushes the worm out or passes through alongside it. Thus finally, the end of the bristle, covered with the posterior portion of the polyp which has been inverted, emerges through the mouth (Plate 11, Fig. 14, *a b*). It is easy to pass the end of the bristle through the mouth because it is wide open. Sometimes, the polyp turns out to be entirely inverted all at once. One can visualize that in this situation the inverted polyp envelops the end of the hog bristle (Fig. 13, *b*) which is now lodged inside it (Fig. 13, *a b*). The exterior surface of the polyp has become the interior touching the bristle, and the interior surface has become the exterior.

When a polyp is out of the water it is difficult to discern its various parts; everything is indistinct. Similarly, one cannot distinguish an inverted polyp which is out of the water at the end of a hog bristle from a polyp that has not been inverted. To further verify the result of my operation, I take the free end of the boar bristle from my right hand into my left hand, and with the right I pick up a magnifying glass. Next I place the polyp into the water of a small glass container prepared in advance. Everything becomes distinct again. By looking at the skin of the polyp, I easily determine whether or not it has been inverted, that is, whether the interior surface of the skin really is on the outside. Usually I have other evidence, however.

It is rare for the polyp to be completely inverted by the movement which causes its inverted posterior end to emerge through its mouth. Thus, when it is put into the water clinging to the bristle, one finds most of the time that its inverted posterior (Fig. 14, *a b*) is emerging from the mouth (Fig. 14, *a*). One also sees, however, that the anterior portion (Fig. 14, *a c*) of the polyp, which terminates in the arms, has not inverted; it is folded over the inverted end. To complete the inversion of this polyp, I take a brush into my right hand while continuing to hold the polyp at the end of a hog bristle submerged in the water. Next, I gently pass the tip of the brush over the uninverted portion (Fig. 14, *a c*) in the direction necessary to invert it, namely from *a* to *c*. This is accomplished in an instant. I can perform this part of the operation using the naked eye. As soon as I have finished, I take up a magnifying glass again and examine the polyp anew. When it turns out to be completely inverted (Fig. 13, *a b*), all that remains to be done is to remove the polyp from the end of the boar bristle (Fig. 13, *d b*). For that purpose, I hold it in the water, I push it gently with the brush from *a* to *b,* and the polyp falls to the bottom of the vessel without being disarranged.

I have done everything that I have just described in the presence of various persons accustomed to making close observations. While I turned the polyp inside out, they kept a continuous watch on it with the aid of magnifying glasses.

Immediately after it has been inverted, the mouth of the polyp closes and its lips even go back inside a little (Fig. 15, *a*). The arms then seem to be joined together in a more or less tight bundle (Fig. 15, *a c*) emerging from the center of the polyp's anterior extremity. Subsequently the lips turn contrariwise outward, however, as if the polyp wished to *evert* itself (if I may be permitted this new expression), that is, to return as it was before being inverted. That is indeed what the polyp does attempt and often succeeds in doing. I have seen polyps which have everted themselves in less than an hour. Others succeeded only after about 24 hours following the operation.

The polyps that do evert fare very well afterwards: they eat, grow, and multiply. In a word, one cannot distinguish them from other polyps.

It is quite extraordinary enough that an animal can be turned inside out and back again not only without dying, but even without appearing to be discomfitted. This did not suffice for me, however. My principal purpose in inverting polyps was to keep them inverted to see whether or not they could survive in the inverted state. I therefore took pains so that they would remain inverted. I succeeded with some by merely depressing their lips with the tip of a brush as soon as they would turn back over the body. This method too rarely succeeds as desired, however, for one to depend on it. When the experiment is to be repeated frequently I have had recourse to a more reliable expedient. It consists of *spitting* the inverted polyp close to its anterior end onto a boar or hog bristle. Here is how I go about it.

After I have successfully inverted a polyp, I put it back into the hollow of my left hand which is partially filled with water, and I examine it again through the magnifying glass. Then with a brush I guide it to the edge of the water and spread it out on my hand as best as I possibly can, and I study its position carefully. Next I take in my right hand a rather thin boar bristle knotted on one side. Holding it close to the tip (Fig. 16, *d*), I drive it into the body of the inverted polyp quite close to the lips. The bristle pierces them and presses against the skin of my hand which is underneath. Then I tip my hand so that the water inside the hollow comes to bathe the polyp which is lifted up and thus becomes the more solidly spitted on the bristle. At that moment I let the opposite end of the bristle (Fig. 16, *c*) gently drop on my hand; picking up the brush again, with its tip I push the polyp toward the middle of the boar bristle (Fig. 16, *a*). When it is there, I pick up the bristle with the polyp and deposit it into a glass vessel (Fig. 16, *f e g h*), taking care that only the ends of the bristle touch the glass (Fig. 16, *c* and *d*). Thus the polyp, kept away from the bottom and the sides of the jar, cannot press against it and exert great efforts to detach itself from the spit in some way or other. I place the knot (Fig. 16, *n*) in the bristle toward the bottom so that if the polyp should slip or fall of its own weight, it would be stopped by the knot and would not be detached from the spit.

No harm comes to the polyp from being spitted. I have tested this in various ways on uninverted polyps. It did not prevent them from eating or multiplying. Neither do the inverted polyps appear to suffer from it.

It is easy to imagine how the boar bristle which crosses their body close to the mouth prevents them from everting.

I have inverted a considerable number of polyps of the second species which have remained inverted and lived for a long time. They ate, grew, and multiplied.

Frequently an inverted polyp that has been spitted in order to prevent it from everting, tears its lips a little and as a result forms two heads. At first the heads have no neck at all, but one develops as the inverted polyp grows. The polyp then resembles those that have been cut lengthwise part way starting at the head.

I have had a number of inverted polyps, each of which had but one head after the operation, each mouth being precisely the same one it had before having been inverted.

I have raised a number of offspring produced by inverted polyps and have seen them multiply.

The first polyps which I inverted had no young at all attached to their bodies, but subsequently I inverted a number that did. After the operation the young end up inside the polyp. If they are already mature, if the area by which they cling to the mother is already quite constricted, then they detach in a short time, that is, at the end of a day or two. During the period prior to separating, they stretch out within the stomach of the inverted mother, and one sees their head and a part of their body emerging through the mother's mouth.

Before the operation I usually separated the young that were already quite mature if they were emerging from mothers that I wished to invert. They are only an inconvenience. The case is not the same, however, with the less mature offspring, that is, those which as yet have no arms or only very short ones and which have not yet constricted at their posterior end. The connecting opening between the stomachs of the young and of their mother is still full size. When the mother is inverted, the young one is able to invert on its own and does. It is precisely as though the fingers of an inverted glove were to invert by themselves. If one attentively examines the body of the mother immediately after she has been inverted, at the spot where one of these young polyps is attached one sees a hollow which fills little by little. Soon after, one clearly discerns the body of the young polyp emerging from it as it inverts itself. I have seen this occur a number of times and with a great deal of pleasure. Only a few minutes are needed for the little polyp to completely invert. Following that, it continues to grow, detaches from its mother, and differs in no way from any other polyp. I have raised such young, and they and their little ones have multiplied in my jars.

As the reader may rightly surmise, the arms of polyps that one is inverting do not invert. Immediately after the polyps are inverted, the arms turn out to be somewhat inside the body whereas beforehand they were somewhat on the outside. In a short time, however, the arms are situated like those of polyps that have not been inverted.

It was while observing a polyp immediately after it had been inverted that I perceived the opening at the base of each arm which I mentioned in the second *Memoir* (p. 77). By means of the channel formed by this opening, the arm communicates with the stomach. One should look for these openings a little below the edges of a newly inverted polyp's mouth. With the aid of a strong magnifying glass, small but very distinct depressions can be seen on the skin. Their location and their number remove any doubt that they are the openings of the polyp's arms. If one becomes accustomed to observing them, one can see these openings gradually grow smaller and finally disappear. The movements and swellings of the skin of the polyp either hide them or close them. Undoubtedly the polyp forms other openings from within the mouth.

The exterior surface of a newly inverted polyp at first is not as smooth as that of a polyp which has not been inverted. As I have described it in the first *Memoir* (p. 30), it is like the interior surface of the polyp's skin. Moreover, most of the time a number of the granules which line this outer surface of an inverted polyp detach, and for several days the exterior of the polyp appears surrounded by detaching granules. Subsequently, however, the surface becomes absolutely as smooth as the exterior of polyps that have not been inverted.

I saw one inverted polyp which ate a small worm two days after the operation. The others did not eat so soon. They remained four or five days more or less without attempting to eat. After that, they all ate as much as polyps that had not been inverted.

I reared one inverted polyp for more than two years. It multiplied prolifically.

I have also inverted polyps of the third species.

As soon as I had succeeded in inverting the polyps, I hastened to perform this experiment in the presence of competent judges. I wanted to be able to refer to other witnesses besides myself to establish the truth of a phenomenon as strange as this experiment revealed. I also made known my desire that others undertake to invert the polyps. At my request, Mr. Allamand at once put his hand to the task and had the same success as I. He inverted a number of polyps so that they remained inverted; they continued to live. He did more: he inverted some polyps that he had already inverted some time previously. Before performing this operation for the second time, he waited until the polyps had eaten after the first inversion. Mr. Allamand also observed them eating after the second operation. Finally, he even inverted one polyp for the third time. It lived for a few days, then died without having eaten. Perhaps, however, its death was not the result of this operation at all.

After Mr. Allamand informed me that he had succeeded in inverting the same polyps twice, I set about duplicating the experiment on two animals, and I had the same success as he did.

Not satisfied with having cut some polyps in different ways and with having inverted others, I wanted to explore whether these different experiments would succeed on the same polyp. I cut a polyp transversely into two; and, when these halves had grown into complete polyps, I inverted them. I saw them eat and multiply after they had been inverted. I performed the same experiment with equal success on halves of a polyp cut into two lengthwise. Finally, in another experiment, I sectioned two inverted polyps which had developed from two halves of a single polyp; I repeat, the parts of these polyps which I first cut, then inverted, and after that cut once again, I saw become complete polyps.

The unusually shaped polyp drawn in Figure 11 of Plate 13 grew from an imperfectly inverted polyp. The operation not having succeeded well, I snipped at this polyp with a scissors, separating it into two parts. At the end of about two months, it was one of these parts which, having consumed a number of meals, exhibited the shape shown in this figure.

I explained earlier that most inverted polyps attempt to evert themselves and in fact can do so. I should add here that all those which evert do not evert completely. This I had occasion to see shortly after having succeeded in inverting some polyps. Afterward I inverted some for the express purpose of observing those which could evert themselves only part way.

As soon as I had seen a few polyps in this extraordinary condition, I was very curious to know what would become of them. Certainly they did not deserve less attention than those which remain completely inverted. These latter polyps (Plate 11, Fig. 16, *a b*) always have a mouth at their anterior extremity. It is the same mouth they had before being inverted, formed by the same lips and bedecked with the same arms; the only difference is that the side of the skin which was outside now is on the inside, and that which was inside is now outside. It is altogether otherwise with a partially everted polyp. After it has been inverted, its lips fold back over the body, and the everted anterior part of the body as well folds back over part of the rest of the inverted body (Fig. 17). The skin of this anterior portion presses down against the other skin and finally fastens to it, forming a sort of cuff (Fig. 17, *a c*) at the anterior extremity of the polyp made up of the everted portion, and inside it, part of the inverted portion. The rest of the body, that is the posterior end (Fig. 17, *a b*) of this extraordinary polyp, is still inverted and stays that way. The lips (Fig. 17, *a*) which formed the mouth of the polyp are no longer at its anterior extremity but turn out to be more or less close to one or the other extremity, depending upon the extent to which the polyp has everted, a highly variable matter. These lips are fastened around the portion of the body that has not everted. The mouth which they form, if it can still be called a mouth, is filled by the portion of non-everted body which passes through it. The arms emerging at its edges are then disposed in a most extraordinary manner. They no longer seem capable of bringing prey to the anterior extremity of the polyp. The direction towards which the arms point varies a great deal, sometimes bearing towards the posterior extremity of the polyp (Plate 11, Fig. 17), but more often bending back towards the anterior end (Fig. 18).

At first I was intent on informing myself fully regarding the condition of these partially everted polyps. The more thoroughly I investigated their strange predicament, the more impatient I became to learn what would become of them. How would they eat? This was the first question I posed for myself. Their anterior extremity (Fig. 17, *c*), formed by the edges of the portion folded back over the body (Fig. 17, *c a*), could remain open as it certainly was shortly after the polyp got into this peculiar state. This opening could serve as the mouth once some arms had grown around its edges. At first this seemed to me the most probable outcome, but I soon had reason to think that nothing of the kind would happen.

Using a strong magnifying glass, I succeeded in observing the anterior end of some inverted polyps which had partially everted a few days previously, and I noted that it was completely closed. It was not closed simply as the mouths of ordinary polyps often are, but closed in a manner to persuade one that it could not open, that in fact no mouth existed. It was not long before I learned how these polyps would be able to eat. Close observation revealed new arms growing near the old ones, and I saw one, then a number of mouths forming near the middle of the body of the polyps, that is, close to the place where the lips had fastened to the body (Plate 11, Fig. 17, *a*). I saw that the lips of these new mouths formed conical protuberances, and it was chiefly by their shape that I recognized them. Here then are these polyps, with one or more mouths formed on their sides. Arms develop around the mouths until, in a word, they look exactly like the mouths of ordinary polyps.

One could perhaps suspect that these mouths forming along the sides of the partially everted polyps are not in truth mouths only, but instead young ones emerging from the sides of the polyps. I venture to affirm that they are not young polyps. I base my statement on continued experimentation, some examples of which I shall present. I have seen some of these new mouths form so close to the old lips of the polyp that one side of their lips was lined with a group of the old arms, and the other side with new arms which had begun to grow. The former were quite long and the others quite short. Finally the new arms grew long and it was not possible then to distinguish them from the others.

A partially everted polyp does not long retain the shape it first has upon getting into this situation (Figs. 17 and 18). It proceeds to take on several other most remarkable shapes, as the reader can judge from the following observations:

I inverted a polyp of the second species on the 26th of September, 1742. This polyp attempted to evert itself. It succeeded in part (Plate 11, Fig. 17). By October 1st the portion that had everted (Fig. 17, *a c*), that is the anterior portion, had already fastened itself to and united with the portion of the body over which it had folded, forming the cuff *a c*. The arms were directed toward the posterior extremity *b*. It seemed to me that no opening was to be found at *c*. By October 4th the positioning of this cuff *a c,* which I shall henceforth call simply the everted portion, had changed in a remarkable way. It no longer lay end to end with the rest of the polyp's body as in Figure 17. The area *a* at which this everted portion clung to the rest of the polyp *a b* had constricted somewhat, and the portion *a c* now formed a right angle with it. Figure 19 illustrates this configuration: *a c* represents the everted portion and *a b* the rest of the body.

On the same day, I observed a polyp's head at the location marked *e* in Figure 19, and I could not comprehend how it had formed. I suspected that it was an offspring, but subsequent observations made me doubt it. This head remained attached to the polyp for a very long period of time, much longer than would an ordinary young polyp. It would have to be viewed either as an unusual young polyp or as a head which had formed at the spot and later, by elongating, had developed a neck. It assumed various shapes which I shall take care to point out. I shall henceforth call it the ambiguous portion.

While carefully examining the polyp used in this experiment through a magnifying glass, on the 4th of October I discovered at location *a* (Plate 11, Fig. 19) several arms that were beginning to grow. They seemed to border one side *a o* of a mouth *a o n* which had formed there. The other side *a n* of the mouth was lined by a portion of the original arms *a d* and *a d* of the polyp. The next day, October 5th, I examined this polyp with great attention, because I was extremely curious to follow the development of the new mouth observed on the previous day. I saw this mouth very distinctly. The lips were shaped into a conical protuberance, and the new arms were shorter than the original ones. I had no more reason to doubt that a mouth had developed approximately in the middle of the body of this polyp, and thus at a location altogether different from where it is usually found. In addition, the mouth was remarkable in that one side of its lips were fringed, as I have said, with new arms, and the other side with some of the old ones. I shall explain shortly what became of the rest of the original arms.

On the same day, October 5th, the everted portion *a c* (Fig. 19), which on the 4th had formed a right angle *c a b* with the non-everted portion *a b,* had drawn closer to the non-everted portion so that the angle between them was now an acute one. This positioning can be seen in Figure 20 where *a c* represents the everted portion and *a b* the non-everted. The ambiguous portion *e* appeared as it had on the 4th. I gave the polyp a worm to determine whether it was already able to use its new mouth, and whether the food that entered through this mouth would spread into the various portions that made up this extraordinary specimen. The worm, falling onto the old arms, was seized at once and carried to the new mouth. It was swallowed just as it would have been by an ordinary polyp. In less than an hour I saw that the portions *a c, a b,* and *a e* (Fig. 20) were swollen, that they were completely filled with the substance of the worm which already had been macerated in the stomach.

Although I observed the polyp in question regularly each day from October 5th to the 9th, I shall not tarry to describe what I saw during this interval because the changes from day to day were not sufficiently noteworthy. On the 9th, however, the polyp was very different from what it had been on the 5th. The difference can be assessed by comparing Figures 20 and 21. The latter shows the polyp drawn life size as it looked on the 9th. The everted portion *a c* and the non-everted portion *a b* are clearly seen to share a common mouth *a*. This is the same mouth I saw beginning to develop on the 4th (Fig. 19, *a o n*), lined partly by new and partly by some of the old arms. The remainder of the old arms can be seen on the everted portion *a c* (Fig. 21) extending from *a* to *t*. On the 9th the ambiguous portion *e* looked as it is portrayed in this figure.

Because the polyp had eaten between the 4th and the 9th of October, it had grown considerably. Between the 9th and the 24th its appearance changed only a little, as can be seen by comparing Figure 21, which depicts it as it was on the 9th, and Figure 22, which shows it as it was on the 24th. It was larger on the 24th. The old arms which on the 9th extended from *a* to *t* (Fig. 21) had disappeared by the 24th. On that day I perceived a head which already had a neck at *u* (Fig. 22). At first I took it to be a young polyp, but it remained the same during the period of more than three months that I observed it, giving reason to believe that it was not an offspring, but simply a head. On the 24th the polyp had two offspring, one sprouting out of the everted portion at *g*, and the other from the inverted portion at *f*. Both detached a few days later.

From the 9th of October until I ceased observing it, I watched this polyp walk a number of times. Ordinarily it would be holding fast by both of its posterior extremities (Fig. 22, *b* and *c*). It is evident that while walking it had to assume postures far different from those shown in Figures 21 and 22. When at rest it was usually disposed more or less as shown in Figure 22, with its two portions *a c* and *a b* drawn together to some extent.

I will not dwell on my observations regarding this polyp during this period; it will suffice to describe its appearance on the date of December 17th. It was clinging to the walls of the jar by its extremities *b* and *c* (Fig. 23), so disposed that the portions *a b* and *a c* were almost end to end. It had four offspring, one sprouting from the non-everted portion *a b* and three from the everted portion *a c*. Another offspring in the shape of a cone (Fig. 23, *k*) was also emerging from the everted section. The ambiguous portion *e* also bore two similar cone-shaped young.

I cared for this polyp until the following February without noticing any substantial changes in it, at which time I ceased observing it. It always had a connecting passage between the everted portion *a c* (Fig. 23) and the non-everted portion *a b*, so that food which entered either by the mouth *a* or the mouth *u* spread throughout both portions. From the end of December, however, I no longer noticed a connection between the ambiguous portion *e* and the rest of the polyp.

The observations I have just recounted suffice to give a sense of how strange these partially everted polyps are. To appreciate the great changes which take place in these animals, the reader need only compare step by step Figures 17, 19, 20, 21, 22, and 23, all of which depict the same specimen. In all of them the letters *a c* indicate the everted portion and *a b* the portion that remained inverted.

These changes are not the same in all polyps. On the contrary, they vary considerably. I have observed a number of partly everted polyps but have never found any two that matched each other exactly as they changed in appearance.

I believe that the following observations made on a partly everted polyp are worth bringing to your attention. When I took up this polyp to turn it inside out, it had two offspring emerging on opposite sides approximately in the middle of its body. Their arms had begun to sprout. Before inverting the mother, I severed the larger of the two young polyps as close to the mother's body as possible. As a result, there was an opening in the mother's skin at the spot where the young one had been. The smaller offspring, which remained attached, ended up in the stomach of the mother when the mother was inverted. Since it was already rather mature, I did not expect to see it invert on its own as I have described the immature offspring doing (p. 159). I anticipated that it would elongate inside the stomach and that its head would emerge through the mouth of the inverted mother, as often happens. Instead, it appeared in the middle of the outside of the mother's body, and on the side opposite to where it had been attached and opposite to which it should have appeared had it inverted on its own. I did not need to search long for the explanation: I saw clearly that after the inversion the young polyp situated inside the mother's stomach had emerged at the opening in the mother's skin made by cutting off the other offspring. Thus it appeared outside the mother in the exact spot where the other young one had been before it was lopped off. This outcome excited my curiosity intensely. We will see speedily now what became of this young polyp.

The mother had been inverted the 24th of September, 1742 and then had partially everted itself. From September 24 until October 4 the everted portion gradually folded over the non-everted portion. By October 4th the two portions were at right angles to each other, as can be seen in Plate 12, Figure 1. Here *a c* represents the everted portion, *a b* the non-everted, and *e i* the offspring described above. The area *a,* at which the portion *a c* held to the portion *a b,* had begun to constrict.

The following day, October 5th, my polyp appeared in a form far different from the one it had the day before. A person who had not previously observed it with great attention and who was not fully accustomed to systematically examining these animals would have had much difficulty in recognizing it. The everted portion *a c* was positioned parallel to the non-everted portion *a b* of Figure 1, and the young polyp *e i* of Figure 1 now lay end to end with the non-everted portion *a b* from Figure 1 and seemed completely united to it. The configuration this partially everted polyp thus manifested is depicted in Figure 2. The everted* portion is shown between the letters *a c* which identify the same portion in Figure 1. The non-everted portion is *e b*. The young polyp *e i* lies end to end with this portion *e b*.

After examining this polyp and marveling at how the young one had grafted itself onto the non-everted section, I gave it some food. Its arms seized a worm which it swallowed through the mouth *i*. The nutrient material spread throughout the body *i e b*, leaving no more room to doubt that the young polyp *i e* was perfectly united to the non-everted portion *e b*, forming a single polyp with it. This specimen was certainly most unusual, composed as it was of an offspring and a portion of its mother onto which it was grafted. What is more, the portion *e b* of the mother (Plate 12, Fig. 2) was inverted whereas the offspring *i e* was not. Thus the posterior portion of this polyp was inside out and the anterior portion was not.

The questions remained, however, of whether the young polyp *i e* would continue to remain united to the portion *e b* of the mother, or whether a constriction would occur at *e* and the portion *i e* would or would not then separate from *e b*. I devoted much attention to this matter. On October 9th the polyp appeared as shown in Plate 12, Figure 3. The young polyp *i e* was still joined to the non-everted* portion *e b*. The old arms which do not border the mouth *a* of the everted portion *a c* can be seen on the body of this portion from *a* to *o*. I fed both the polyp *i e b* and the everted portion *a c*, and I noted that there no longer existed any connection between them. The polyp *a c* detached from the polyp *i e b* on October 15th. I continued to nurture both polyps for two months following their separation. They multiplied and performed all the functions of complete polyps. The portion *i e* of the polyp *i e b* was originally the offspring *i e* shown in Figure 1. This anterior portion never did separate from the posterior portion *e b* (Fig. 3), which was originally the non-everted portion *a b* (Fig. 1) of the polyp that had been inverted on September 24th.

---

*See Errata, Book I, p. 60.

The polyp I shall be describing now was inverted on October 3, 1742. By the next day it had partly everted (Plate 11, Fig. 18). On October 10th I clearly saw that it had formed three heads near the area (Fig. 18, *a*) where the lips had stopped after the polyp had partly everted. On that day these three heads already had necks. The polyp looked as it is drawn in Plate 12, Figure 4. The portion that remained non-everted is designated *a b,* the same as in Figure 18, Plate 11. The rest of the polyp *a d c g e* (Plate 12, Fig. 4) developed from the everted portion. The three heads which formed at *a* of Figure 18, Plate 11 are designated *d, g, e.*

Between October 10th and 24th I fed this polyp a number of times, and it took in food through its three mouths (Plate 12, Fig. 4, *d, g, e*). During this interval it grew and changed its appearance. Plate 12, Figure 5 shows it as it appeared on the 24th. By comparing it with Figure 4, one can readily discern the change that occurred. The various portions of the polyp are designated by the same letters in both figures. The inverted portion is *a b,* and the everted portion is *a d c g e,* with the heads at *d, g, e.* In Figure 5 these heads are designated together with their necks by the letters *a d, n g, n e.* The letter *c* in both figures marks the anterior end of the polyp after it had everted. This spot is designated by the same letter in Figure 4 and in Figure 18 of Plate 11.

Next, the portion *c o* (Fig. 5) split into two from *c* up to *o,* so that by November 6th the part of the polyp marked *b a c d* was attached to the other, marked *c n g e o,* only by a small process at *o.* This configuration can be seen in Figure 6 which shows the polyp as it appeared on December 6th with the two portions *b a c d* and *o c n g e* still holding together at the process *o.* The portion remaining non-everted is *a b;* the part of the everted* portion, marked *a o c* in Figure 5, is found here at *a c.* One of the heads with its neck is shown at *a d.* It no longer holds the position it did on October 24th. Instead of forming the angle *d a b* (Fig. 5) with the section *a b,* it now forms with *a b* almost a straight line *d a b* (Fig. 6). The portion *o c* in Figure 6 designates the other part of *a o c,* Figure 5, which had split from *o* to *c,* as I have mentioned. These two portions *a c* and *o c* of Figure 6 developed from the split parts, growing larger as did the rest of the polyp between the 24th of October and the 6th of November as a result of the food it consumed. The two other heads of this polyp are shown at *n g* and *n e.*

The two portions *b a d c* and *c n g e o* separated on the 10th of November, appearing as shown in Figures 7 and 8 of Plate 12. The section *a d* (Fig. 7) no longer formed a straight line with *a b* as on November 6th, but rather an

_____

*See Errata, Book I, p. 60.

angle *b a d* as it had on October 24th. The two may be compared by looking at *b a c d* of Figure 7 and *b a c d* of Figure 5. The portion *c n g e* (Fig. 8) was almost unchanged from the 6th of November to the 10th and was not much different from what it had been on October 24th, as can be seen by comparing Figures 5, 6, and 8. After November 10th I still kept the two parts shown in Figures 7 and 8 for some time. Their appearance did not change much.

While observing polyps that were partially everted I often wondered what would become of the inverted portion of the polyp that was covered by the everted portion. It occurred to me that perhaps, when the surfaces of the two portions were set against each other, they would fuse so that henceforth they would form but a single skin. Accordingly, the inverted portion would act as a sort of lining to the everted portion (Plate 11, Figs. 17 and 18, *a b*) which had covered it over. I conjectured on this basis that if the everted portion were lined by the inverted portion which it covered, then a polyp could perhaps be lined by inserting another polyp inside its body. The object then was to place a polyp into the stomach of another polyp so that the walls of the stomach of the first would be pressed against the external surface of the skin of the second polyp. It would then be a matter of observing whether or not the skins of these two polyps would fasten together so that the *inner* polyp would act as a lining to the *outer,* and whether or not they would form but a single polyp afterward.

Immediately I applied myself to searching for a method of inserting one polyp into another. Here is the one that seemed the most efficacious to me: First, I feed the polyps chosen for this experiment so that the food they take will make them swell and thereby assist in keeping their mouths open when this becomes necessary. Next, I place the polyp that is to be inserted into the stomach of the other polyp upon my left hand and out of the water; I make it contract as much as possible. I press it with a brush so that the food in its stomach emerges part way through its mouth, forcing the mouth to open. Then, I take a boar bristle with my right hand, insert the thicker end into the polyp's mouth, and push the bristle to the further end of its stomach. It is even relatively easy to insert the end of a boar bristle into the body of a polyp that is not full of food. Next, I place upon my hand the polyp into which the first is to be inserted. I make it open its mouth and then insert into its stomach both the first polyp and the boar bristle at the end of which it is fastened. The end of the boar bristle thus comes to be covered by two polyps placed one inside the other. I dip it into the water of a small jar arranged in advance, slide the polyps off onto the bottom, and examine them through a magnifying glass.

At first I took no precautions to force the inner polyp to remain inside the outer; but once having noticed that it was coming out, I took care thereafter to spit both polyps after one was inside the other in such a way that both were pierced through by the same boar bristle.

I expected to see the two polyps placed one inside the other adhere to each other and result in a single polyp, but I saw at once a completely different outcome, one that I would hardly have imagined possible. The polyps, placed one inside the other, although spitted together as I have said, separated a number of times; the inner polyp emerged from the outer. The experiments that I am going to report will reveal the remarkable manner in which this separation occurs.

On October 17, 1742, I inserted a polyp of the second species into another polyp of the same species. Beforehand I had sliced off a small section from the posterior extremity of the outer polyp in order to make it shorter than the inner one and to leave an opening in that area out of which the posterior extremity of the inner polyp could emerge. Thus I could ascertain that the body of the inner polyp extended from one end to the other of the stomach of the outer polyp. Next I thrust a single boar bristle through both these two polyps and placed them in a jar. They looked as is depicted in Plate 12, Figure 9, where *c* marks the head of the inner polyp, *a* the head of the outer, *d* the posterior extremity of the inner polyp, and *b* that of the outer. One sees from this figure that the head and tail of the inner polyp emerge from the body of the outer one. The same day, October 17, the posterior extremity *b d* (Fig. 9) of the inner polyp *c a b d* tore through the skin of the outer polyp; it passed through this cleft so that by evening it already had reached the spot marked *i* in Figure 10. Thus the portion *i d* of the inner polyp already had emerged from the outer polyp. The opening made in the portion *i b* of the outer polyp always closed up as soon as the portion *i d* moved forward. I never saw this opening, not even a scar.

On the morning of October 18th, I found that the posterior end of the inner polyp had continued to emerge from the outer by tearing it along one side and that it was already outside up to the spot marked *e* (Fig. 11) where the boar bristle held both polyps transfixed. In the evening I found that the inner polyp was emerging from the outer a little above *e*. One could see clearly that the boar bristle pierced through the polyps one after the other, as depicted in Figure 11. The inner polyp is identified by the letters *c a i d,* the outer by *a i b*. The letters *i d* mark the portion of the inner polyp that had emerged from the portion *i b* of the outer polyp by gradually cleaving through it.

On the 21st the anterior end *c a* (Plate 12, Fig. 11) of the inner polyp began to split the lips of the outer polyp. By the next day the portion *c a i* (Fig. 11) of the inner polyp had emerged completely from the portion *a i* (Fig. 11) of the outer polyp. The two polyps now lay side by side (Fig. 12). Thus, the polyp *c e d* which had been inserted into the polyp *a e b* on the 17th of October had emerged from it in four days by gradually rending the outer polyp from *b* to *e* and from *a* to *e*.

I saw each of these two polyps eat a worm on the 25th of October. The next day I pulled out the boar bristle. The polyps were slightly attached at *e* (Fig. 12), but a few days later they separated completely. I kept them for some time and in no way did they differ from other polyps.

The two polyps I shall now discuss also had been inserted into one another. Again the inner of the two emerged from the outer but in a manner somewhat different from that described above. In this case, I placed one inside the other on October 17, 1742 and pierced them through with a bristle a little closer to the head than in the preceding experiment. On the day of the operation the posterior end *b d* (Fig. 13) of the inner polyp protruded from the posterior extremity *b* of the outer,* but the head of the inner polyp barely emerged from the outer polyp's mouth. On the following day, October 18, only the arms of the inner polyp remained visible (Fig. 13, *a o*), and later even the arms completely reentered the body of the outer polyp. On the morning of the 19th, however, I found the arms of the inner polyp once more protruding from the mouth of the outer polyp. The mouth *a* had assumed a conical shape. Around noon I saw the head itself of the inner polyp outside the mouth of the outer polyp. By four o'clock I could no longer see anything but the arms. So it remained all that evening, during which I observed the polyps a number of times. At nine-thirty in the evening I counted five arms (Fig. 13, *a o*) of the inner polyp emerging from the mouth *a* of the outer polyp.

At eight-thirty on the morning of October 20th, a few arms of the inner polyp protruded from the mouth of the outer polyp. At two o'clock in the afternoon, the entire head emerged. From that moment until nine-thirty in the evening I consistently saw the same state of affairs. On the 21st I perceived only a few arms of the inner polyp protruding from the mouth of the outer. But at midday I found a great alteration. The tail of the inner polyp

---

*See Errata, Book I, p. 60.

had partially broken through one side of the outer polyp and was emerging from it! By the evening the portion *e d* (Fig. 14) of the inner polyp, which was emerging from the outer polyp *a e b*, already was nearing point *e* where the boar bristle *n f* pierced through the two polyps. Neither the head nor the arms of the inner polyp appeared outside the mouth *a* of the outer polyp, although five arms *t o, t o*, etc. emerged through the opening *t* made by the boar bristle. This opening lay at the side of the polyp opposite that through which the portion *e d* of the inner polyp was emerging.

On October 22nd I saw only two arms *t o, t o* (Plate 12, Fig. 15) outside of the opening *t* where five had been protruding the day before. Another of those arms *e o*, however, was emerging through an opening *e* on the opposite side. It seemed to me that I even saw the head of the inner polyp show a little at this latter opening. Finally the head did emerge on the 23rd through the opening *e*, and the inner polyp ended up outside and alongside the outer polyp as in Figure 12.

Thus we see that in the first experiment the inner polyp had rent the portion *a e* of the outer polyp to have its anterior end emerge. On the other hand, in the second experiment the inner polyp withdrew its head into the body of the outer and then pushed it out through the side of the outer polyp together with its entire anterior section *c e* (Fig. 12).

Here I need to note that the inner polyp remained for four days inside the outer polyp and at the end of that time emerged alive and healthy. It is this phenomenon to which I referred in the second *Memoir* (p. 71).

It could be proposed, perhaps, that the inner polyp was not digested by the outer polyp, because the boar bristle that transfixed the outer polyp prevented it from digesting food as would a polyp that was not pierced through. To counter this argument I need only remark that this particular outer polyp, which had been unable in four days to kill the inner one, ate several worms while it was spitted. All of them died and were digested in its stomach in a short time.

The third experiment that I now wish to recount began on October 9th, 1742. The two polyps that were placed one inside the other looked like those portrayed in Plate 12, Figure 9. The outcome of this experiment was very different from that of the previous two that I have just reported. These two polyps had not altered their position by the 12th of October, on which day I drew out the bristle that held them pierced together. That evening, however, I noticed that the extremity *c a* (Plate 13, Fig. 1) of the inner polyp, which was protruding from the outer polyp's mouth *a*, was much longer than on the preceding days. I do not know whether that was the case because the extremity *c a* had elongated more, or because the inner polyp was in fact emerging further out of the outer polyp's mouth. On the 13th day of October, the tail portion of the inner polyp tore through the outer polyp's body from the bottom upward to the area marked *i* (Fig. 2). The portion *i d* marks the tail of the inner polyp which has broken through the outer one. On the 16th they remained as they are shown in Figure 2 of Plate 13.

I fed a worm to the inner polyp *c a i d*, the only one of the two polyps which could eat because the outer polyp's mouth *a* was filled with the part of the inner polyp that was emerging from it. I noticed, however, that the food spread not only through the inner polyp which had eaten it, but also through the outer. This fact was easily confirmed by the swelling of the posterior end *i b* (Fig. 2) of the outer polyp *a i b*. If the food had spread only through the inner polyp *c a i d*, then only the portions marked *c a*, *a i*, and *i d* would have swollen. The portion *a i*, because it was a part of the outer polyp *a i b*, would have appeared inflated only because it contained the portion *a i* of the inner polyp hidden underneath it. Figure 3 shows the polyps as they appeared some time after the inner polyp had eaten. It is easy to observe that the portion *i b*, which forms the outer polyp's posterior end, has swollen and has filled with food.

In order for the food to have passed from the inner to the outer polyp, the portion *a i* (Plate 13, Figs. 2 and 3) of the inner polyp must have been open somewhere.

By the 19th of October the anterior end (Fig. 2, *c a*) of the inner polyp had changed its position very markedly. On that day it no longer emerged from the outer polyp's mouth at *a*, but having torn through the lips of the outer polyp, it protruded from its body a little below the head. This new configuration is shown in Figure 4, where the letters *c o* designate the anterior end of the inner polyp, marked *c a* in Figure 2.

Since the mouth (Fig. 4, *a*) of the outer polyp was now clear, I attempted to feed it a worm. The polyp swallowed it and after the worm had been digested, its nutrient juices spread perceptibly throughout all the parts of both polyps (Fig. 5, *a i*, *c o*, *i b*, and *i d*). I saw the same thing happen on the 22nd after I had fed a worm to the portion *c o*. The food also passed from this portion into the other sections of the polyp. Thus there was not any reason to doubt that the inner polyp, comprised partly of sections *c o* and *i d* (Fig. 4), had formed a connecting passage to the outer polyp *a i b*, that is to say, the inner polyp was open between *i* and *o*. It is impossible for me to describe the condition of the portion *a i* of the inner polyp which was concealed within the outer shortly after the opening developed. But I can attest that it subsequently merged with the portion *a i* (Fig. 4) of the outer polyp, without presuming to explain, however, how this occurred.

For about three months I continued to examine the portion *a i* from many perspectives, and it always appeared to me like a portion of an ordinary polyp. The portions *c o* and *i d* of the inner polyp seemed as though they were grafted onto the outer polyp *a i b*. They were connected to it as are young polyps with their mother. These two united polyps grew and multiplied, as indicated in Figure 6 of Plate 13 which shows them as they were the 6th of November. At that time they were sprouting four offspring *e, e, e, e,* and previously a few young had already detached from them. The anterior portion *c o* of the inner polyp seemed farther away from the head of the outer polyp than it had been on October 19th (Fig. 4). I had every reason to believe that this difference was not the result of its changing its place, but rather the result of an elongation of the portion *a o* (Fig. 6) of the outer polyp. The portion *c o* remained open to the portion *a i* until the 20th of November. Food plainly passed from one into the other. Then the portion *c o* constricted at point *n* (Fig. 7) and broke away at that point from the remainder of the animal on December 24. After that it was a separate polyp (Fig. 8), as was the other part of the two formerly united polyps (Fig. 9). The posterior end *i d* of the inner polyp (Fig. 9) was united completely with the outer polyp *a i b* and continued to be so united until February, 1743 when I thought it no longer necessary to continue observing it.

This last experiment which I have just reported indicates that in a way one polyp can be grafted on to another. It has shown us at the least that the portions *c o* and *i d* (Fig. 6) of the inner polyp remained united for a long time to the outer polyp *a i b,* as a bough or a slip is united to the stock to which it has been fastened.

This possibility of grafting was not the only question I was pursuing when I started to insert polyps into each other. I wanted to see whether the inner polyp would fuse with the outer without emerging from it. Would the inner act as a sort of lining, the two polyps forming a single animal?

Of all the experiments I performed with this goal in mind, none fulfilled my expectations better than the one I started on October 22nd, 1742. The polyp which I inserted into another on that day never emerged from it, either in whole or in part. I had inverted the polyp before inserting it; and, as usual, I carefully spitted the two polyps together after one was inserted inside the other. I cannot explain what became of the body of the inner polyp, whether it was dissolved in the stomach of the outer polyp, or whether the inner body merged with the outer. I can definitely state, however, that I could still see the body of the inner polyp lying inside the outer polyp several days after it had been inserted. As for the head of the inner polyp, I was certain that it had fused with that of the outer polyp. The lips of the outer polyp adhered to the neck of the inner polyp, and after a certain time the two heads of these polyps formed but one which had two rows of arms (Plate 13, Fig. 10, *a* and *c*). On various occasions, while feeding the polyp *c a b* (Fig. 10), I saw clearly that it had but one mouth, the one belonging to the inner polyp. The mouth of the outer polyp *a* was filled by the inner polyp's head; or rather, it no longer was a mouth. Only the arms which had lined its lips could be distinguished. They formed the second row of arms (Fig. 10, *a*) that I just mentioned.

I raised this polyp from the 22nd of October, 1742 until the middle of February, 1743 when it died of sickness. During this time it grew and multiplied and I was always able to distinguish the two rows of arms on its head.

This specimen (Fig. 10) belongs among the more extraordinary polyps that we have been discussing till now. It was composed of an outer and an inner polyp or, at least, the anterior portion of the inner polyp which could easily have become a complete polyp on its own. Thus it is certain that it was composed of two polyps, and that being the case, this experiment can be viewed as the opposite of the first one we performed on the polyps. The first [on regeneration] taught us that from one polyp two could be made; and this one teaches that two polyps can be made to form one.

It was by chance that I learned to reunite severed parts of polyps. I had placed into a small glass container two pieces of polyps which I had used in an experiment. Examining them the next day, I noticed that they were attached to each other. I cut them anew and purposely placed them next to each other. Once again they joined together. I needed no additional incentive to stimulate me to perform this experiment carefully. Here is the way I go about reuniting parts of polyps.

After sectioning a polyp, I place the two halves that I wish to reunite on the bottom of a shallow glass containing water only about four to five millimeters in height. Pushing them with a tip of a brush, I bring them together and arrange them so that the two ends at which they are to unite are touching, or nearly so. If they are touching, I watch them through a magnifying glass to discern whether they continue to touch. If they are not quite touching, I examine whether or not the two halves, by elongating, bring their ends together until they meet. Frequently the parts become displaced, and their ends, which need to touch, pull away from one another. In that case I use the brush to place them again in the requisite position. If I see that their ends continue to touch, however, I take great care not to disturb them even slightly.

After the ends have remained touching for some time, a quarter hour, a half hour, an hour, the observer begins to be aware that they are clinging together. This does not occur with all parts which are in contact for a while. Often one finds that the pieces of polyp draw apart afterwards either because they contract, or because their positions become disarranged. In that event, it is necessary to rearrange them in the proper position. It seemed to me helpful to force the parts to contract before bringing them together. This procedure widens the two ends which must touch, and the wider they are the more likely the reunion. Such, at least, is my conjecture.

Although I take care to bring the parts of polyps together immediately after they have been sectioned, this precaution is not absolutely necessary in order for the experiment to succeed. I have seen some pieces which only began to join two hours after the polyp had been cut. Perhaps if one persisted in bringing the pieces back into contact whenever they drew apart, one would finally succeed in making them all reunite.

As for me, I exercised my patience in this regard only to a limited degree. I continued to bring the pieces of polyps back together as they became disarranged only for an hour or two. Generally this kind of experiment is successful less often than those I have discussed until now.

At first the two pieces of polyps unite only by a portion of their extremities. The polyp they form is tightly constricted at the area where they have joined (Plate 13, Fig. 12, c). As the polyp is fed more and more, the constriction lessens and finally disappears entirely.

I shall now describe a few particular polyps, the pieces of which I reunited, so that the examples may provide a clearer idea of the process.

On November 5th, 1742, at two o'clock in the afternoon, I sectioned a polyp of the second species transversely. I brought the two halves together (Fig. 12, a c and c b) so that the anterior end of the second part touched the posterior end of the first. A little later they appeared to be beginning to rejoin. In the evening I saw clearly that they were attached to each other. As I mentioned earlier, however, the polyp that they formed was tightly constricted in the area where the reunion occurred (Fig. 12, c). The next day I gave the polyp a worm. My purpose was to learn not only whether or not it would eat, but more especially whether a connecting passage existed between the two portions. I had evidence of this, for I saw a part of the worm pass from the anterior half (Fig. 12, *a c*), which had swallowed it, into the posterior half *c b,* which had united with the anterior portion. Both portions then became swollen (Fig. 13, *a c* and *c b*), and the place at which they had reunited (Fig. 13, *c*) appeared still quite constricted. On the 15th of November the polyp started to sprout a young one and then a number of others. By the 20th the constriction was no longer visible.

After succeeding in reuniting portions of the same polyp, I undertook to unite those taken from different polyps. On November 7th I sectioned two polyps of the second species. I brought the first part of one in contact with the second part of the other, and vice versa. The experiment succeeded with only two of these parts, that is, the first half of one polyp fastened onto the second half of another. There was a remarkable aspect to this union in that the first part was whitish and the second of a rather dark brown color. At eleven o'clock in the evening of the same day, November 7th, I gave this polyp a worm. The following morning at eight o'clock I found that the worm had passed completely into the second portion which was quite swollen whereas the first portion was not. I gave the anterior portion something to eat.

The spot at which these two joined halves touched each other began to widen only on the 13th of November. Towards the end of the month after the polyp had consumed some meals, the constriction was no longer noticeable, and both halves displayed the same coloring. They were brown. This polyp, composed of portions of two different polyps, then multiplied both above and below the area where the portions were joined; that is, each portion produced young. I continued to observe it until the end of the following February.

I have also attempted to hold parts of polyps of different species together. I cannot say that I succeeded. I saw only two pieces of polyps, one from the second and the other from the third species, which remained slightly attached for about two weeks before separating again. I have repeated this experiment neither with sufficient care nor with sufficient frequency, however, to assert that it cannot succeed.

When I had succeeded in reuniting parts of polyps, grafting them so to speak, I immediately reported the outcome of the experiment, as was my custom, to Mr. Réaumur. The phenomenon was not new to him. He kindly informed me of his experience in a letter dated December 14th, 1742, in which he stated: "After I had thrown polyps which had been cut in half into a sharply tapered glass, one on top of the other, two pieces rejoined. This gave me the idea of attempting on a larger animal what you have accomplished on a small one. I contemplated joining together the halves of two different sea anemones sectioned at an angle perpendicular to their base. Mr. de Villars, a doctor at La Rochelle, was so kind as to undertake these operations. He had to resort to sutures in order to secure the two halves one against the other. A few of his experiments succeeded, but many more failed. He is still pursuing them at this time."

As we see, this excerpt from Mr. Réaumur's letter not only confirms what I have said about the reuniting of polyps, but in addition informs us that this experiment has also succeeded on very different animals.

The similarity that exists in this respect between sea anemones and freshwater polyps with arms shaped like horns provides further proof that the strange properties encountered in polyps are also found in other animals. This I have already shown in regard to the polyp's manner of multiplying by shoots and as a result of being sectioned (*Memoir* III, p. 126 and following). Indeed, as various facts of their natural history were introduced in the course of these *Memoirs,* I could have discussed additional similarities to be found between polyps and other animals. I did not wish to interrupt the account of my experiments too often, however, and I chose to postpone until now the discussion that seems necessary to me on this subject.

For the creatures whose natural history I have just presented the name polyp is fitting, because, as seen earlier (*Memoir* I, p. 27), they are similar in appearance to marine polyps [that is, squids, cuttlefish, and octopuses]. It is only natural to ask whether these animals resemble each other in other ways, and especially whether the marine polyps have the same unusual characteristics which are now known to exist in the freshwater polyps with arms shaped like horns. Insofar as I am able, I shall try to answer these questions by reporting some of what is to be found in the literature on the marine polyps.

Marine polyps are large animals in comparison to those studied in these *Memoirs.* According to reports, it seems that they commonly are between one and three feet in length. Pliny (*Naturalis Historia,* Vol. 9, Chap. 30) spoke of a monstrous polyp with arms thirty feet long and so thick that they almost exceeded a man's grasp. I believe one may be allowed to doubt the accuracy of such a description.

Marine polyps are carnivorous, and like freshwater polyps, they use their arms to seize prey and carry it to their mouth.

I believe there is nothing written on the anatomy of these animals more detailed or exact than the works of Swammerdam (*Biblia Naturae,* p. 875 and following). This skillful naturalist has carried out an anatomical study of the cuttlefish, which is a species of polyp. His description reveals that its inner structure is not so simple as that which the freshwater polyp with arms shaped like horns appears to us to have.

Naturalists unanimously agree that marine polyps are male and female, that they mate and are oviparous. (See especially Aristotle, *Hist. Animal.,* Vol. 5, Chap. 18; Massarius in Pliny, Vol. 9, p. 182 and following, Basel 1537.) I have personally examined some eggs of the cuttlefish. They are black and a number of them cling together in a shape like a bunch of grapes. I do not know whether marine polyps have natural means of reproduction other than that.

The most interesting question here is whether they can be multiplied by cutting them into sections. Aldrovande has said (prologue in volume on insects, p.17) that the polyp cut into pieces continues to live, and I do not know that there is anything further on this subject by other authors. It is quite clear that the statement means only that the pieces continue to give signs of life, not that with time each regenerates its missing parts to become a complete animal. We will be able to learn whether or not they possess this capacity only by sectioning the marine polyps and systematically observing the separated parts.

It has been mentioned, however, that regeneration occurs in certain parts of a marine polyp's body, as it commonly does in those of freshwater polyps with arms shaped like horns. The reference is to their arms, whether completely or partially missing. Elian (Aelian, *Hist.* Vol. 1, Chap. 17) states that marine polyps eat their own arms when they lack food. Other authors deny this claim (Aristotle, *Hist. Animal.,* Vol. 8, Chap. 2; Pliny, *Hist. Nat.,* Vol. 9, Chap. 29), asserting that they are gnawed off by conger eels. All agree, however, that the arms grow back.

I am citing these various authors without attempting to judge what degree of faith one should place in the points they advance. In order for such judgments to be valid, one would need to know what observations gave rise to their conclusions and how these observations were made. These details are not to be found in the writings of any of these authors.

In the third *Memoir* (p. 126) I indicated the similarities that exist between the polyps with arms shaped like horns and the tufted polyps. There is no need to dwell on them now.

There are a number of species of other animals, smaller by far than the tufted polyps, which it seems also ought to be classified among the polyps. Mr. Réaumur has found great quantities of them around Paris and in Poitou. A number of species also abound in the ditches around The Hague and in those of Sorgvliet. All these animals have a mouth at one of their extremities, and their arms, or what they use for arms, are also located at the same extremity. Several of the species known to me are certainly carnivorous. Observing them through a strong magnifying glass, one clearly sees small animals drawn into their mouths. It would be difficult to give a good description of these animals without the aid of a great number of illustrations. Leeuwenhoek described one of these species in the letter which we have already discussed (*Philosophical Transactions,* number 283, Article 4, p. 1305). They are shown greatly magnified in the plate of the *Philosophical Transactions* (Fig. 8 N W V and I T S) which bears on the article cited.

Regarding the different species of polyps which can afford us abundant material for new discoveries, I must say a few words about the find made by Mr. Hughes, an English clergyman. In the water of a grotto on the island of Barbados, he saw some organisms which he first mistook for flowers, but which he later believed should be classified as animals. When he attempted to pick these supposed flowers, they hid themselves, as we have said the tufted polyps do. After several minutes they reappeared and gradually unfolded themselves. The observations of Mr. Hughes suggest that the rays, or rather arms, which surround the anterior end of these animals are used to seize various small creatures that swim in the water. The *Philosophical Transactions* contain a more ample description of his observations on this subject. It is proper to note that Mr. Hughes, at the time he made these observations, had no knowledge of the discoveries on the polyps made in Europe a few years since. In fact, his observations were made first.

I have said that it is not known whether marine polyps can multiply after they are cut into sections. I should add here that this attribute does not appear to have been known in any animal, at least by the naturalists whose writings have come down to us. They have indeed mentioned various animals which, after being cut into pieces, continue to show signs of life for some time. Aristotle declares (*Hist. Animal.,* Vol. 4, Chap. 7) that most small creatures live after being sectioned. He says nothing, however, which would indicate that he had observed any regeneration taking place whereby each part developed into a complete animal. He appears to have noted only that some parts survived longer than others. He reports that animals which have long bodies and many legs live the longest after being sectioned, and that the parts continue to move after they have been separated, one frontward, the other backward. On this subject he cites the example of the centipedes. St. Augustine (*Lib. de Quantitate Animae*) also reports an example of an animal of this kind, that is a species of millepede. The one he saw was cut into a number of pieces; "each one," he says, "moved about in such a manner that had we not separated them ourselves, and had the wounds not been visible, we would have mistaken them for that many different animals."

It does not seem, then, that any experiments were done to see what would become of the pieces of animal in which some movement remained after sectioning. People in various localities certainly have believed that pieces of this or that animal continued to live and even develop into complete animals; in truth, however, these ideas were hardly passed on except as opinions of the common people. In the preface to the sixth volume of his *Memoirs* on the insects (pp. 61 and 62), Mr. Réaumur presented what fishermen of Normandy had observed in regard to the starfish. I shall shortly report what Swedish peasants have said about another aquatic animal and how their beliefs were viewed.

Earlier in the present *Memoir* we saw that, when brought into contact, pieces of a sectioned freshwater polyp can reunite and form a single animal. The same ability was attributed to lizards by Jean Baptiste Porta (*Phytognomonica,* Vol. V, Chap. 12, Edit. Francof.) It appears that he derived this passage from Elian (Aelian, *Hist.,* Vol. 2, Chap. 23) who says the same thing of lizards and serpents. Even today this ability is rather generally attributed to these same animals. I cannot say whether this opinion is based on observations or on conjectures. In essence, our knowledge of other properties quite common to animals has prepared us for this phenomenon. Experience has taught that in many situations, parts of animals, when they adjoin each other, can reunite as easily as do parts of plants. Doctors and surgeons have encountered this many times in regards to the human body. In fact, precautions must be taken in cases where it is desirable to prevent such a rejoining from occurring. Thus it is not astonishing that the same capability should exist in little creatures. There is no reason to be surprised that parts of a cut animal continue to live for a time after being separated, or that separated pieces of others can each develop into a complete animal, or that pieces of these animals can reunite and then form a single animal.

I have already mentioned that the ancients spoke of the regeneration of the arms of marine polyps. I will add here that they spoke also of a regeneration which occurred in other animals, that is, in the tails of lizards and serpents. Both Aristotle (*Hist. Animal.,* Vol. 2, Chap. 17) and Pliny (*Nat. Hist.,* Vol. 9, Chap. 29) maintain that the tail grows back. This view is still generally accepted today as regards the lizards. If the reader should wish to be informed of facts that are thoroughly proven and thoroughly detailed on the regeneration of parts in certain animals, however, he should read Mr. Réaumur's elegant *Memoir* (Mém. de l'Acad., 1712, p. 226) on the discoveries he has made in this regard upon the crayfish.

Because the polyps which have been the subject of these *Memoirs* became well known principally at first for their characteristic of being able to increase in number as a result of their being sectioned, it is not surprising that there was a tendency to regard them as plants, this attribute having been recognized only in plants until then. This impression would be intensified all the more if one next considered their other properties that we have revealed. Indeed, who would not believe that they are plants and not animals if, after learning that polyps can be multiplied by sectioning them, one heard that they multiply naturally by sending out shoots and that they can be grafted? In considering them plants, we would be but following the generally recognized rules regarding the nature of plants and animals. But to truly follow these same rules, we could not deny either that these polyps are animals when we observe their locomotion and especially when we see them seize small creatures and carry them to their mouth with their arms, swallow them, and digest them.

But what are we to make of the polyps if we are guided only by the rules based on the properties of plants and animals known to date, when we consider their ability to be turned inside out without dying and without even causing any noticeable change in them? Since this attribute is completely unknown, either in plants or in animals if I am not mistaken, one then would have to conclude that in this respect the polyps are neither plant nor animal. Thus, if one were to cling scrupulously to the quite generally accepted ideas on the nature of plants and animals, it would follow that a polyp, in view of its various properties, is at the same time plant and animal, and neither animal nor plant.

From this we can judge how little certainty there is to these supposed general rules. Independent of the evidence furnished us by the polyps, it is easy to see that we do not know enough about plants and animals to formulate general rules about their nature. Let me explain in regard to animals.

When I say that they are not sufficiently known to us, I mean that only a limited number of them are known and these not in sufficient depth.

I believe I can safely pose here as a principle that there are animals unknown to us. Then I ask, can we formulate general rules which would apply to these unknown animals? To be able to do so, one would need to be certain that a perfect similarity exists between the unknown animals and those that are known to us, something no one would know how to prove. To the contrary, the diversity of attributes that are found among the known animals seems to imply that as yet unimagined traits exist in the unknown ones.

Let us assume that at first man knew only the animals which do not undergo any metamorphoses. Would it then have been reasonable to say that animals in general are unable to pass through various stages? And, based on this rule, when he began to learn of organisms that underwent transformations, would he have been justified in concluding that they were not animals because they went through such changes?

Could there not be between the animals we know and those we do not know differences as great or even greater than those between animals that undergo metamorphoses and those that do not?

It is much more reasonable not to judge the unknown ones strictly by the animals we know, because from all indications the number of the unknown greatly exceeds the number of those known to us.

Not only can we not formulate general rules applicable to unknown animals, but we are as yet far from being able to make infallible rules respecting the animals we know, because most of them are known to us only very superficially. Moreover, it would be most rash to judge the characteristics of lesser known animals, whether in whole or in part, by the characteristics of those we understand better.

Experiences repeated every day teach us clearly that a large number of animals, the quadrupeds, for example, cannot be multiplied by means of cutting them into a number of pieces. When they are sectioned, each piece dies. One should conclude from these experiences only that the severed parts of certain animals cannot develop into complete animals and that, according to all indications, this conclusion holds true for a number of other animals that have not been sectioned. This limited inference, however, has been made into a general rule, and it is supposed that no animal can be multiplied by cutting it into sections. Nevertheless, several known animals, and among them the frequently encountered earthworms, possess this attribute which seems incompatible with the nature of animals. How many presumably well-known animals have attributes equally extraordinary and just as unexpected?

It would be equally easy to prove that we do not know plants sufficiently well to be able to formulate general rules regarding their nature.

Because we are still far from knowing all the characteristics that plants and animals may possess, it is not surprising that those characteristics which are distinctive for each of these classes of organism are not very well understood.

At first, before an attempt has been made to delve deeply into the nature of plants and animals, nothing seems easier than to find the characteristics that distinguish one class from the other. Unless I am mistaken, this comes from making judgments only on the basis of narrow ideas instead of comparing general ideas, that is the abstract idea of a plant with the abstract idea of an animal.

It is very easy to distinguish a horse from a oak tree, and it is customary to think that it is no more difficult to distinguish an animal from a plant. These are, however, very different matters. To come to this conclusion, try separating from the idea of a horse and that of an oak all the attributes that are particular to this species of animal and this species of plant, keeping only those that are common to all known animals and all known plants. The more the ideas one fashions are general, the more the particular attributes which help differentiate a horse from an oak tree disappear; and the more one approaches the general idea of an animal and that of a plant, the less one will find differences between them.

Accordingly, naturalists who have applied themselves to discovering the distinctive characteristics of plants and of animals have encountered great difficulties. They have found very difficult a matter which appears very simple to those who have not studied the subject.

No more respected authority can be cited at this juncture than the renowned Boerhaave. What effort has this great man not expended in studying plants and animals! Nonetheless it seems that he has found but a single general and essential difference between these two classes of organisms. This difference, which is set forth in the beginning of his *Elementa*

Chymiae in the articles where he deals with plants and animals (p. 57 and following, Leiden Edition 1732), consists of the manner in which plants and animals draw in their nourishment. "The nourishment of plants," says Mr. Boerhaave, "is drawn in through external roots, that of animals through internal roots. The external part, called a root, which draws nourishment from the substance in which it is situated, is sufficient to distinguish a plant from any animal known until now." And in the definition that he gives of an animal body, Mr. Boerhaave says, "it contains within it vessels instead of roots through which it draws the food material." Subsequently, after further dwelling on the comparison of a plant and an animal, he adds: "in this the similarity and the difference which exists between a vegetable and an animal may be found."

It is useful to note that Mr. Boerhaave does not conclude that the distinction exists between absolutely all plants and animals. He remains cautious and says only that "this external part, called a root, which draws nourishment from the substance in which it is situated, is sufficient to distinguish a plant from any animal" known until now.

What we have said in the second *Memoir* about the manner in which the polyps feed proves that they are animals according to the definition of Mr. Boerhaave. Independent of the authority of this accomplished man, I doubt that many persons would deny that polyps should be classified this way.

These are animals, nevertheless, which share more similarities with certain plants than do many other animals known to us.

For this reason we could, perhaps, be influenced to call them animal-plants. I do not know whether this name should be used except in a manner of speaking because it seems to me that there are fewer reasons to classify polyps definitively as animal-plants than as simple animals. In order to be able to decide that a particular organism is neither plant nor animal but belongs to some intermediate class, it would be necessary to know precisely all the attributes of plants and animals. As seen above, we are still very far from such knowledge. Only when we succeed in acquiring it can we create other classifications of organisms. In the meantime, it is much more natural to consider the polyps and various other organisms which have been given the name of zoophytes as animals which show more noteworthy similarities to plants than do other animals.

We have not believed we should undertake to explain, either in whole or in part, the extraordinary phenomena we have reported. It is too dangerous in the subject of natural history to abandon experience and allow the imagination to lead us. In following such a path, one risks arriving only at uncertain hypotheses which can become detrimental to the progress of this science should one have the misfortune to become prejudiced in their favor. Instead of clarifying phenomena through new experiments, we have recourse to a hypothesis, or rather a prejudgment, which not only spares us the trouble of observing, but which often serves but to compound our errors.

Thus it is, for example, that for so many centuries it was believed that a great number of animals arose from the decay of the bodies in which or on which they were found. This was the origin attributed to all small creatures whose manner of multiplying was not known. Not that one could not have learned about it, even with moderate effort. But prejudice is blinding and hinders one from even thinking about investigating. As soon as this preconception was doubted and observations begun, those animals were seen to lay eggs or give birth to young like so many others. These facts that philosophers were unable to see, or rather whose prejudice either prevented their searching for or seeing perhaps when they were under their very eyes, could be discovered by children amusing themselves for a time by watching the small creatures, as I had the pleasure of experiencing with children a short time ago.

It is quite probable that were it not for a number of preconceptions that have gained currency, natural history would be more advanced than it presently is. This view holds especially true in regard to the multiplication of animals by sectioning. Had this process been presumed possible, it would long since have been a property recognized in a number of animals. In fact, everything seemed ripe for the making of this discovery. That diverse well-known animals did not die at once after being cut into two or into a number of sections had been witnessed and wondered about for a long time. Both life and movement had been noted in the separated parts. Why then did we not think to preserve the pieces and to observe them to see whether each one would, by regenerating the missing portions, grow into a complete animal? The answer is because it was presumed impossible.

Mr. Linnaeus, professor of botany at Upsala University, reports an example in an abstract printed at the beginning of a speech which he gave in 1743. He says that peasants from the province of Smaland and from other areas unanimously insist that if the worm which he, Linnaeus, has named *Gordius,* is cut into a great number of pieces, each retains the ability to move; when put back into the water, each grows back a head, a body, or a tail. Mr. Linnaeus then adds that blinded by their adopted principles, naturalists considered what was said about this worm a fantasy so ridiculous and contrary to nature that they did not undertake a single experiment to verify it.

Notice that it was only uneducated people who could not have been imbued with the prejudices of the schools, who quite simply believed based on the facts they observed that parts of an animal can become complete animals.

The idea then current that no animal could multiply by being sectioned, as we found occurred with polyps, seems likely only to have made us miss opportunities for discovering that property earlier in other animals. By a quite peculiar chance, however, that idea contributed greatly to this discovery. For I would not have undertaken the experiment that led to this result had I not assumed that the pieces of an animal could not become complete animals. I refer the reader to the beginning of the first *Memoir* (p. 5 and following).

As soon as we learned that polyps can multiply by being sectioned, we were cured of the long-held presumption in this regard as to the nature of animals. Attempts were made to section various species and to observe the resulting parts. As mentioned earlier (*Memoir* I, p. 3), in a short time a number of animals were found to possess the same attribute as the polyps.

There is reason to hope that these discoveries will produce a number of good results. They should naturally jolt us into a great suspicion of those general rules within which, if I may say so, we have attempted to confine Nature. Such rules can only serve as obstacles to our knowledge. Moreover, the discoveries should help us realize that what we know is quite insignificant when compared to the prodigious number of marvels encompassed by Nature. We still know too few parts of the admirable Whole which is the Work of a Being infinite in all respects. What little we know of the parts is not enough for us to be able to explain all the facts presented to us.

In order to extend our knowledge of natural history, we must put our efforts into discovering as many facts as possible. If we knew all the facts that Nature holds, we would have the explanation of them, and we would see the Whole which these assembled facts fashion. The more we know of them, the more we will be in a position to delve deeply into some parts of this Whole. Thus we cannot work better to explain the facts we know than by trying to discover new ones. Nature must be explained by Nature and not by our own views. These are too limited to envision so grand a Design in all its immensity. The beauty of Nature certainly shines forth all the more when what we know about it is not mixed with our fancies. Seen clearly, Nature inspires within us ideas more worthy of the infinite wisdom of its Author and thereby more suitable for shaping our spirits and our hearts. This thought is what we should keep before us in all our researches.

EXPLANATION OF
THE FIGURES OF THE FOURTH MEMOIR
PLATE 11

FIGURES 1 and 2 represent the halves *a c* and *c b* of a polyp of the second species which has been cut in two. Figure 1 shows the first part: *a* is the head, *c* the place at which it has been cut. This end *c* often appears open shortly after the sectioning. Figure 2 depicts the second part with its anterior end *c* open.

The second part *c b* is shown also in Figure 3, but its anterior end *c* is closed and a little swollen.

Figure 4 shows a second part *c b*. The arms are beginning to sprout at the anterior end *c*.

Figure 5 pictures the lips of a polyp which have been cut away with a scissors from the rest of the body. From these lips the polyp's arms emerge.

Figure 6 shows a piece of the lips of a polyp from which two arms are emerging.

Figure 7 represents the half *a e i b* of a polyp which has been cut lengthwise. This part has not yet closed. Its interior surface is shown. This half has four arms at its anterior extremity *a e*.

In Figure 8, half of a polyp cut lengthwise is shown as such halves often appear shortly after having been sectioned. Such portions coil around themselves starting at one of their extremities. The one pictured here has started at its anterior end. The arms are inside the coil.

Figure 9 also represents half *a e b i* of a polyp cut lengthwise. Its edges *a i* and *e b* have begun to draw together and to join at the posterior end *i b*. They already have joined as far as *c*. No scar is discernible in the area where these edges have joined.

Figure 10 depicts the same half as is shown in Figure 9, but with its edges entirely joined together. At this stage, this half already can be considered a complete polyp. Two new arms can be seen beginning to grow in the spaces between the original set.

Figure 11 portrays a Hydra with seven heads.

Figure 12 shows a polyp *a b*, quite contracted and quite swollen, as it appears when placed on the hand outside the water in order to invert it. The extremities *c* and *e* of a larva of the midge [*Chironomus;* see Book I, p. 48] that is bent over double are emerging from the mouth of the polyp, thus holding it wide open. The boar bristle *b d*, which is used to invert the polyp, is placed in such a way that one of its extremities touches the posterior end of the polyp at *b*. When the boar bristle is set in this manner, one can begin to push against the end of the polyp at *b* to make it enter the body in the process of inverting the polyp.

In Figure 13 an inverted polyp *a b* is shown situated at the end of the boar bristle *b d* as it is immediately after the operation is completed.

PLATE 12

FIGURE 1 represents a partly everted polyp. The everted portion is marked *a c*. The letters *a b* indicate the posterior part of the polyp which has remained inverted. Immediately after the inversion the young polyp *e i* was inside the stomach of the mother. It has emerged, however, through the hole left in the skin of the mother made by cutting off another young polyp which also had been growing from the mother's body.

In Figure 2 the same polyp is shown as in the first figure, but after it has undergone quite a considerable change. The everted portion is *a c*, the portion remaining inverted is *e b*. The young polyp *e i* has grafted itself on to the inverted portion *e b*.

Figure 3 depicts the same polyp as it appeared 4 days later. From one side of the everted portion *a c* emerge some of the arms the polyp had prior to its inversion. The inverted portion is *e b*, and the young polyp is *e i*.

Figure 4 shows a different partly everted polyp. The posterior portion of the polyp which has remained inverted is marked *a b*. A considerable change has occurred in the everted portion *a d c g e:* the polyp has developed three heads *d, g, e*.

Figure 5 represents the same polyp as does Figure 4, only 14 days later. The portion which has remained inverted is *a b;* the everted portion extends through *a d c n g e o*. The three heads, designated *d g e* in the preceding figure, are shown here in Figure 5 with their necks as *a d, n g*, and *n e*.

Figure 6 shows the same polyp as it appeared 13 days after it was in the condition depicted in Figure 5. The portion marked *o c* in Figure 5 has split from *o* to *c*, and the two portions *b a c d o* and *c n g e o* in Figure 6 have remained attached to each other only by a small process *o*. The portion which has remained inverted is marked *a b*. The two portions *a c* and *o c* were indicated united as *o c* in Figure 5. The three heads and their necks *a d, n g*, and *n e* are designated by the same letters as in Figure 5.

The two portions which were shown in Figure 6 attached only at the process *o* are shown detached in Figures 7 and 8. In Figure 7, *a b* marks the inverted portion and *a d* one of the heads. The other heads *n g* and *n e* are seen in Figure 8.

Figure 9 represents two polyps, one of which has been inserted inside the other. The outer polyp is *a b*, the inner polyp *c a b d*. A hog bristle *n f* pierces through both these two polyps at *e*.

The same polyps are depicted in Figure 10. The outer polyp is marked *a i b*, the inner polyp *c a i d*. The portion *i d* of the inner polyp has broken through the portion *i b* of the outer polyp and has emerged from it in this way. The boar bristle which pierces these two polyps is marked *n f*.

In Figure 11 the same polyps are shown. Now, however, portion *i d* of the inner polyp has torn further through the portion *i b* of the outer polyp and has emerged from it at a point beyond the area marked *e* where the hog bristle *n f* originally had pierced both polyps together. After they assumed the shape shown in this figure, the bristle pierced first through the inner polyp and then the outer. The outer polyp is *a i b*, and the inner is *c a i d*, and the boar bristle piercing these two polyps is *n f*.

Figure 12 depicts the two polyps after the inner one has split the lips of the outer polyp to emerge completely from it. The outer polyp is marked *a e b*, the inner is *c e d*, and the hog bristle which pierces them through is *n f*.

Figure 14 also represents a polyp *c a b* at the end of a boar bristle *b d*, but this one has not been completely inverted yet. Having been pushed inside by the boar bristle, the posterior portion *a b* has been inverted, and in this inverted condition, is emerging through the mouth *a*. The anterior portion *a c* has remained non-inverted.

Figure 15 is of an inverted polyp *a b* as such polyps usually appear shortly after they have been removed from the end of the boar bristle where they were lodged. At first the lips are turned inward and the arms are joined together more or less tightly into a bundle that appears to emerge from the middle of the anterior end of the polyp.

Figure 16 depicts an inverted polyp *a b* pierced through close to its head *a* by a boar bristle *c d* knotted at *n*. It is placed in a jar in such a way that only the ends of the bristle touch the glass; that is, extremity *d* touches the rim *f h*, and extremity *c* touches the bottom *e g*. The knot *n* is under the polyp *a b* so that if the polyp should slip from *a* toward *n*, it would be stopped by the knot *n* and would not come off the bristle by slipping all the way to *c*.

Figure 17 depicts a partly everted polyp *c a b*. That is to say, after having been completely inverted, as is *a b* in Figure 15, its lips have turned over onto the outside of the inverted body as the whole anterior portion *a c* then does. The mouth of the polyp before it partially everted is at *a* (Fig. 15). The everted portion, or cuff, is marked *a c*. The letters *a b* designate that portion of the polyp which has remained inverted and which is still visible. The arms of this polyp point towards the posterior end *b*.

The arms shown in Figure 18, which also depicts a [partially] everted polyp *c a b*, bend back towards *c*, which is the new anterior end of this polyp.

Figure 19 illustrates the same polyp as does Figure 17. The everted portion *a c*, which in Figure 17 was end to end with the non-everted portion *a b*, in Figure 19 has bent over and arranged itself in such a way that the two portions have formed a right angle *c o b* with each other. The previously formed arms are marked *a d, a d*, a few of which emerge from the old lips *a n* of the polyp. New arms are starting to grow at *a o*. The new mouth *a o n* is lined on one side *a n* by the old arms and on the other side *a o* by new ones. The inverted portion is *o b*. The letter *e* indicates a portion of the body [the ambiguous portion], the origin of which I do not know.

The same polyp is shown in Figure 20, one day older than in Figure 19. The letters *a c* mark the everted portion; *a b* the inverted portion; *e* the ambiguous portion; *a d, a d* the old arms; *a i* the new arms; and *a* the mouth, designated as *a o n* in the preceding figure.

In Figure 21 the same polyp is shown after having eaten, grown, and altered its shape quite markedly. The everted portion is *a c;* the inverted portion is *a b;* the ambiguous portion is *e*. The remaining old arms that were not used to line the new mouth, formed at *a*, extend from *a* to *t*.

Figure 22 shows the same polyp once again, only larger and older. The letters *a c* mark the everted portion; *a b* the inverted portion; and *e* the ambiguous portion. A new mouth formed at *u* without my having noticed it. At *a* is the mouth designated *a o n* in Figure 19.

Figure 23 still depicts the same polyp, but shown after it has assumed an aspect far different from that in the two preceding figures and with a number of young on its body. The everted portion is at *a c*, the inverted at *a b*. From the ambiguous portion *e* two conical shaped offspring have emerged. Another conical offspring *k* has emerged from the portion *a c*. The letter *a* is the mouth designated as *a o n* in Figure 19. The mouth *u* is marked the same as in Figure 22.

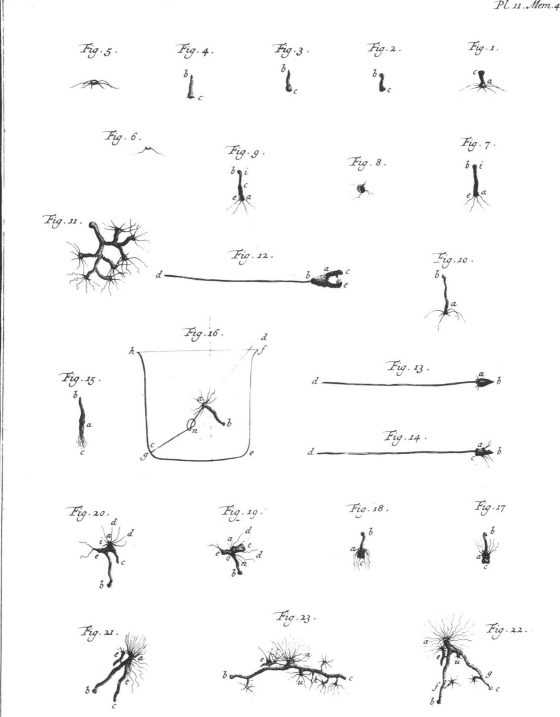

Pl. 11. Mem. 4.

Fig. 5.

Fig. 4.

Fig. 3.

Fig. 2.

Fig. 1.

Fig. 6.

Fig. 9.

Fig. 8.

Fig. 7.

Fig. 11.

Fig. 12.

Fig. 10.

Fig. 16.

Fig. 15.

Fig. 13.

Fig. 14.

Fig. 20.

Fig. 19.

Fig. 18.

Fig. 17.

Fig. 21.

Fig. 23.

Fig. 22.

P. Lyonet delin. et sculp. 1744.

PLATE 13

FIGURE 1 represents two polyps placed one inside the other. The outer polyp is *a b,* the inner one *c a b d.*

In Figure 2 the same polyps are shown, but the portion *i d* of the inner polyp has emerged from the outer polyp by tearing through the portion *i b.* The outer polyp is *a i b,* the inner *c a i d.*

Figure 3 depicts the polyps of Figure 2, but after they have eaten. It is the inner polyp that has swallowed a worm through its mouth *c.* The worm then has filled all parts of both polyps, making them swell. The outer polyp is *a i b,* the inner *c a i d.*

Figure 4 depicts the same polyps, but after the portion marked *c a* in Figure 2 of the inner polyp has torn through the lips of the outer polyp, assuming the position shown in Figure 4. The portion designated *c o* in Figure 4 is the one that is marked *c a* in Figure 2. The outer polyp is *a i b,* the inner *c o i d.*

Figure 5 shows the polyps of Figure 4, but swollen with the food which has been introduced into their stomachs through the mouth *a* of the outer polyp. The food has spread into the outer polyp *a i b* and the inner *c o i d,* which is grafted onto the outer polyp.

After having eaten a number of times these polyps grew and multiplied. Figure 6 shows the outer polyp *a i b,* the inner *c o i d,* and the young polyps *e, e, e, e.*

The same polyps are depicted again in Figure 7. Portion *c o* of the inner polyp has become tightly constricted at *n.* The outer polyp is *a i b.* The inner is *c n o i d.*

Figure 8 represents the portion *c n* of Figure 7 which has separated from the rest [of the grafted polyps] at *n.*

Figure 9 shows the remainder of those two grafted polyps. The outer polyp is *a i b.* The letters *o n* mark the remnant of the anterior portion of the inner polyp designated *c n o* in Figure 7. The posterior portion of the inner polyp is *i d.*

Figure 10 depicts a polyp into which I put another one [that I had previously inverted]. The head of the inner polyp united with the head of the outer. The two heads then merged in such a way that they could no longer be recognized except by the two rows of arms one above the other at *a* and *c.*

Figure 11 represents a half of a polyp that I tore while inverting it. After that I snipped at it a few times with the scissors. At the end of two months it turned out as is depicted in this figure.

Figure 12 shows two halves of polyps which have been divided and which are already rejoined a little. The first portion is *a c,* the second portion *c b.* The place where they have rejoined, *c,* is quite constricted.

The same two halves appear in Figure 13, but swollen with food. The first portion is *a c,* the second is *c b,* and the area where they have rejoined, which is quite constricted, is at *c.*

THE   END

192

Figure 13 represents two other polyps, one of which is inserted into the other. The outer polyp is *a b*. The posterior extremity of the inner polyp is *b d*. The rest of the body of the inner polyp is hidden within that of the outer polyp except for five arms *a o, a o, a o, a o, a o* protruding from the mouth *a* of the outer polyp. The boar bristle which pierces through both polyps is *n f*.

Figure 14 shows the same polyps. The inner polyp is *t e d*. Its arms *t o, t o, t o, t o, t o,* have emerged through an opening which was at *t*. The letters *n f* designate the hog bristle which runs through both these polyps.

Figure 15 again shows the two polyps of Figure 13. The outer polyp is marked *a t b*, the inner one *t e d*. Two of the arms *t o, t o* of the inner polyp emerge through the opening *t*. Another arm of the same polyp is at *e o*.

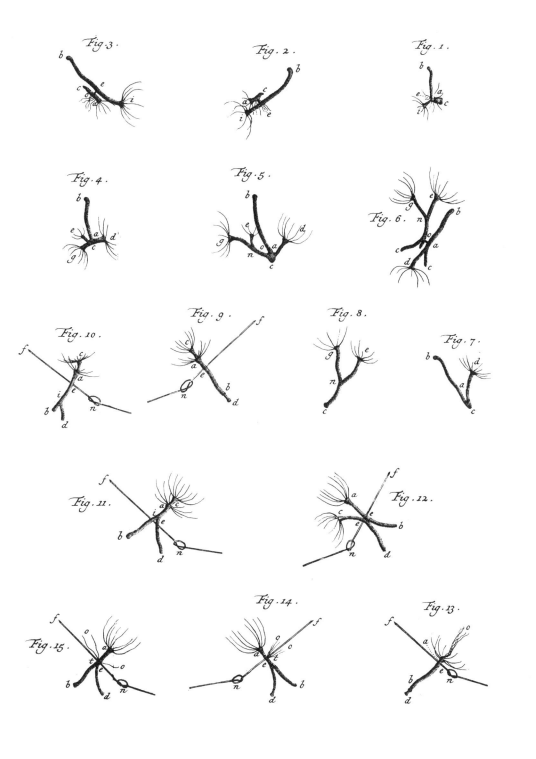

Pl. 12. Mem. 4.

P. Lyonet delin. et sculp. 1744.

Pl. 13. Mem. 4.

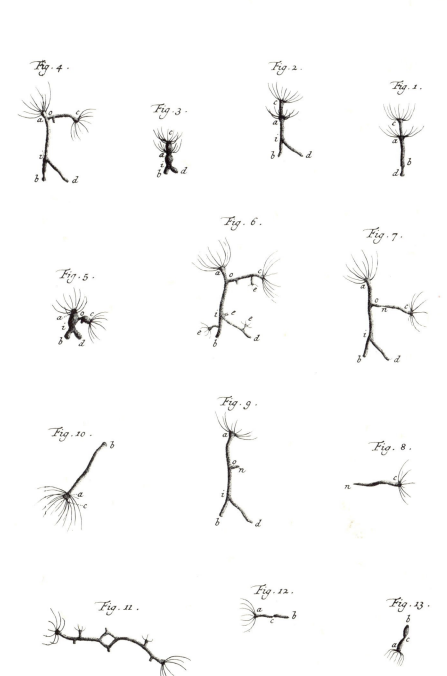